COBIT® 5

for Risk

COBIT® 5

AN ISACA® FRAMEWORK

ISACA®

With more than 110,000 constituents in 180 countries, ISACA (*www.isaca.org*) helps business and IT leaders maximize value and manage risk related to information and technology. Founded in 1969, the non-profit, independent ISACA is an advocate for professionals involved in information security, assurance, risk management and governance. These professionals rely on ISACA as the trusted source for information and technology knowledge, community, standards and certification. The association, which has 200 chapters worldwide, advances and validates business-critical skills and knowledge through the globally respected Certified Information Systems Auditor® (CISA®), Certified Information Security Manager® (CISM®), Certified in the Governance of Enterprise IT® (CGEIT®) and Certified in Risk and Information Systems Control™ (CRISC™) credentials. ISACA also developed and continually updates COBIT®, a business framework that helps enterprises in all industries and geographies govern and manage their information and technology.

Disclaimer

ISACA has designed and created *COBIT® 5 for Risk* (the 'Work') primarily as an educational resource for assurance professionals. ISACA makes no claim that use of any of the Work will assure a successful outcome. The Work should not be considered inclusive of all proper information, procedures and tests or exclusive of other information, procedures and tests that are reasonably directed to obtaining the same results. In determining the propriety of any specific information, procedure or test, assurance professionals should apply their own professional judgement to the specific circumstances presented by the particular systems or information technology environment.

Reservation of Rights

ISACA

3701 Algonquin Road, Suite 1010
Rolling Meadows, IL 60008 USA
Phone: +1.847.253.1545
Fax: +1.847.253.1443
Email: *info@isaca.org*
Web site: *www.isaca.org*

Provide Feedback: *www.isaca.org/cobit*
Participate in the ISACA Knowledge Center: *www.isaca.org/knowledge-center*
Follow ISACA on Twitter: *https://twitter.com/ISACANews*
Join ISACA on LinkedIn: ISACA (Official), *http://linkd.in/ISACAOfficial*
Like ISACA on Facebook: *www.facebook.com/ISACAHQ*

COBIT® 5 for Risk
ISBN: 978-1-60420-457-5
2

ACKNOWLEDGEMENTS

ISACA wishes to recognise:

COBIT for Risk Task Force
Steven A. Babb, CGEIT, CRISC, ITIL, Betfair, UK, Chairman
Evelyn Anton, CISA, CISM, CGEIT, CRISC, Uruguay
Jean-Louis Bleicher, CRISC, France
Derek Oliver, Ph.D., CISA, CISM, CRISC, FBCS, FISM, MInstISP, Ravenswood Consultants Ltd., UK
Steve Reznik, CISA, ADP Inc., USA
Gladys Rouissi, CISA, ANZ Bank, Australia
Alok Tuteja, CGEIT, CRISC, Mazrui Holdings LLC, UAE

Development Team
Floris Ampe, CISA, CGEIT, CRISC, CIA, ISO 27000, PwC, Belgium
Stefanie Grijp, PwC, Belgium
Bart Peeters, CISA, PwC, Belgium
Dirk Steuperaert, CISA, CGEIT, CRISC, ITIL, IT In Balance BVBA, Belgium
Sven Van Hoorebeeck, PwC, Belgium

Workshop Participants
Elza Adams, CISA, CISSP, PMP, IBM, USA
Yalcin Gerek, CISA, CGEIT, CRISC, TAC, Turkey
Jimmy Heschl, CISA, CISM, CGEIT, Bwin.party Digital Entertainment Plc, Austria
Mike Hughes, CISA, CGEIT, CRISC, 123 Consultants GRC Ltd., UK
Jack Jones, CISA, CISM, CRISC, CISSP, Risk Management Insight, USA
Andre Pitkowski, CGEIT, CRISC, APIT Informatica Ltd, Brazil
Eduardo Ritegno, CISA, CRISC, Banco de la Nacion Argentina, Argentina
Robert Stroud, CGEIT, CRISC, CA Technologies, USA
Nicky Tiesenga, CISA, CISM, CGEIT, CRISC, IBM, USA

Expert Reviewers
Elza Adams, CISA, CISSP, PMP, IBM, USA
Mark Adler, CISA, CISM, CGEIT, CRISC, CIA, CRP, CFE, CISSP, Wal-Mart Stores Inc., USA
Michael Berardi, CISA, CGEIT, CRISC, Bank of America, USA
Peter R. Bitterli, CISA, CISM, CGEIT, Bitterli Consulting AG, Switzerland
Sushil Chatterji, CGEIT, CMC, CEA, Edutech Enterprises, Singapore
Frank Cindrich, CGEIT, CIPP/G, CIPP/US, Deloitte and Touche, LLP, USA
Diego Patricio del Hoyo, Westpac Banking Corporation, Australia
Michael Dickson, CISA, CISM, CRISC, CPA, GBQ Partners, USA
AnnMarie DonVito, CISA, CISSP, ISSAP, ISO 27001 Lead Auditor, PRINCE2 Practitioner,
     ITIL Foundation V3, Deloitte AG, Switzerland
Ken Doughty, CISA, CRISC, CRMA, ANZ, Australia
Urs Fischer, CRISC, CISA, CPA (Swiss), Fischer IT GRC Consulting and Training, Switzerland
Shawna Flanders, CISA, CISM, CRISC, CSSGB, PSCU, USA
Joseph Fodor, CISA, CPA, Ernst and Young LLP, USA
Yalcin Gerek, CISA, CGEIT, CRISC, TAC, Turkey
J. Winston Hayden, CISA, CISM, CGEIT, CRISC, South Africa
Mike Hughes, CISA, CGEIT, CRISC, 123 Consultants GRC Ltd., UK
Duc Huynh, CISA, ANZ Wealth, Australia
Monica Jain, CGEIT, Southern California Edison (SCE), USA
Waleed Khalid, CISA, MetLife, UK
John W. Lainhart, IV, CISA, CISM, CGEIT, CRISC, CIPP/G, CIPP/US, IBM Global Business Services, USA
Debbie Lew, CISA, CRISC, Ernst and Young LLP, USA
Marcia Maggiore, CISA, CRISC, Consultor en TI, Argentina
Lucio Augusto Molina Focazzio, CISA, CISM, CRISC, Independent Consultant, Colombia
Anthony Noble, CISA, Viacom Inc., USA
Abdul Rafeq, CISA, CGEIT, A.Rafeq and Associates, India
Salomon Rico, CISA, CISM, CGEIT, Deloitte, Mexico
Eduardo Ritegno, CISA, CRISC, Banco de la Nacion Argentina, Argentina

ACKNOWLEDGEMENTS (CONT.)

Expert Reviewers (cont.)
Paras Kesharichand Shah, CISA, CGEIT, CRISC, Vital Interacts, Australia
Mark Stacey, CISA, FCA, BG Group plc, UK
Robert Stroud, CGEIT, CRISC, CA Technologies, USA
Greet Volders, CGEIT, Voquals N.V., Belgium
John A. Wheeler, CRISC, Gartner, USA
Tichaona Zororo, CISA, CISM, CGEIT, CRISC, CIA, CRMA, EGIT | Enterprise Governance of IT (PTY) LTD,
    South Africa

ISACA Board of Directors
Tony Hayes, CGEIT, AFCHSE, CHE, FACS, FCPA, FIIA, Queensland Government, Australia, International President
Allan Boardman, CISA, CISM, CGEIT, CRISC, ACA, CA (SA), CISSP, Morgan Stanley, UK, Vice President
Juan Luis Carselle, CISA, CGEIT, CRISC, RadioShack, Mexico, Vice President
Ramses Gallego, CISM, CGEIT, CCSK, CISSP, SCPM, Six Sigma Black Belt, Dell, Spain, Vice President
Theresa Grafenstine, CISA, CGEIT, CRISC, CGAP, CGMA, CIA, CPA, US House of Representatives, USA, Vice President
Vittal Raj, CISA, CISM, CGEIT, CFE. CIA, CISSP, FCA, Kumar and Raj, India, Vice President
Jeff Spivey, CRISC, CPP, PSP, Security Risk Management Inc., USA, Vice President
Marc Vael, Ph.D., CISA, CISM, CGEIT, CRISC, CISSP, Valuendo, Belgium, Vice President
Gregory T. Grocholski, CISA, The Dow Chemical Co., USA, Past International President
Kenneth L. Vander Wal, CISA, CPA, Ernst and Young LLP (retired), USA, Past International President
Christos K. Dimitriadis, Ph.D., CISA, CISM, CRISC, INTRALOT S.A., Greece, Director
Krysten McCabe, CISA, The Home Depot, USA, Director
Jo Stewart-Rattray, CISA, CISM, CGEIT, CRISC, CSEPS, BRM Holdich, Australia, Director

Knowledge Board
Christos K. Dimitriadis, Ph.D., CISA, CISM, CRISC, INTRALOT S.A., Greece, Chairman
Rosemary M. Amato, CISA, CMA, CPA, Deloitte Touche Tohmatsu Ltd., The Netherlands
Steven A. Babb, CGEIT, CRISC, Betfair, UK
Thomas E. Borton, CISA, CISM, CRISC, CISSP, Cost Plus, USA
Phil J. Lageschulte, CGEIT, CPA, KPMG LLP, USA
Anthony P. Noble, CISA, Viacom, USA
Jamie Pasfield, CGEIT, ITIL V3, MSP, PRINCE2, Pfizer, UK

Framework Committee
Steven A. Babb, CGEIT, CRISC, Betfair, UK, Chairman
David Cau, France
Sushil Chatterji, Edutech Enterprises, Singapore
Frank Cindrich, CGEIT, CIPP/G, CIPP/US, Deloitte and Touche, LLP, USA
Joanne T. De Vita De Palma, The Ardent Group, USA
Jimmy Heschl, CISA, CISM, CGEIT, Bwin.party Digital Entertainment Plc, Austria
Katherine McIntosh, CISA, Central Hudson Gas and Electric Corp., USA
Andre Pitkowski, CGEIT, CRISC, APIT Informatica, Brasil
Paras Kesharichand Shah, CISA, CGEIT, CRISC, Vital Interacts, Australia

Special recognition for financial support:
Los Angeles Chapter

TABLE OF CONTENTS

LIST OF FIGURES

EXECUTIVE SUMMARY

Introduction

Information is a key resource for all enterprises. From the time information is created to the moment it is destroyed, technology plays a significant role in containing, distributing and analysing information. Technology is increasingly advanced and has become pervasive in enterprises and the social, public and business environments.

COBIT 5 provides a comprehensive framework that assists enterprises in achieving their objectives for the governance and management of enterprise information technology (IT). Simply stated, COBIT 5 helps enterprises to create optimal value from IT by maintaining a balance between realising benefits and optimising risk levels and resource use. COBIT 5 enables IT to be governed and managed in a holistic manner for the entire enterprise, taking into account the full end-to-end business and IT functional areas of responsibility and considering the IT-related interests of internal and external stakeholders.

COBIT 5 for Risk, highlighted in **figure 1**, builds on the COBIT 5 framework by focusing on risk and providing more detailed and practical guidance for risk professionals and other interested parties at all levels of the enterprise.

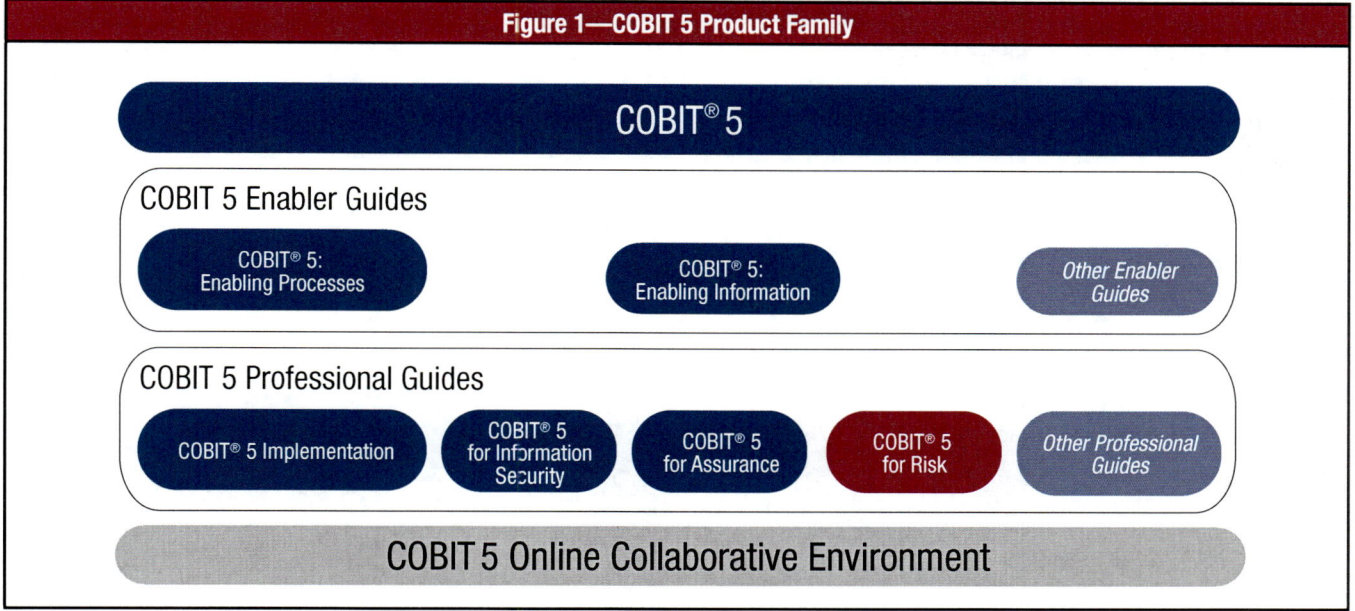

Figure 1—COBIT 5 Product Family

Terminology

COBIT 5 for Risk discusses IT-related risk. Section 1, chapter 2 defines what is meant by IT-related risk; however, for ease of reading, the term 'risk' is used throughout the publication, which refers to IT-related risk. The guidance and principles that are explained throughout this publication are applicable to any type of enterprise, whether it operates in a commercial or non-commercial context, in the private or the public sector, as a small, medium or large enterprise.

COBIT 5 for Risk presents two perspectives on how to use COBIT 5 in a risk context: risk function and risk management. The **risk function perspective** focuses on what is needed to build and sustain the risk function within an enterprise. The **risk management perspective** focuses on the core risk governance and management processes of how to optimise risk and how to identify, analyse, respond to and report on risk on a daily basis. These perspectives are explained in detail in section 1, chapter 2. Risk; section 2A, The Risk Function Perspective; and section 2B, The Risk Management Perspective and Using COBIT 5 Enablers.

Drivers for Risk Management

The main drivers for risk management in its different forms include the need to improve business outcomes, decision making and overall strategy by providing:
• Stakeholders with substantiated and consistent opinions on the current state of risk throughout the enterprise
• Guidance on how to manage the risk to levels within the risk appetite of the enterprise

- Guidance on how to set up the appropriate risk culture for the enterprise
- Wherever possible, quantitative risk assessments that enable stakeholders to consider the cost of mitigation and the required resources against the loss exposure

To that purpose, this publication:
- Provides guidance on how to use the COBIT 5 framework to establish the risk governance and management functions for the enterprise
- Provides guidance and a structured approach on how to use the COBIT 5 principles to govern and manage IT risk
- Demonstrates how *COBIT 5 for Risk* aligns with other relevant standards

Benefits of Using This Publication

Using *COBIT 5 for Risk* increases the enterprise risk-related capabilities, which provide benefits such as:
- More accurate identification of risk and measurement of success in addressing that risk
- Better understanding of risk impact on the enterprise
- End-to-end guidance on how to manage risk, including an extensive set of measures
- Knowledge of how to capitalise on investments related to IT risk-management practices
- Understanding of how effective IT risk management optimises value, with business process effectiveness and efficiency, improved quality and reduced waste and costs
- Opportunities to integrate IT risk management with enterprise risk and compliance structures
- Improved communication and understanding amongst all internal and external stakeholders due to the common and sustainable globally accepted framework and language for assessing and responding to risk
- Promotion of risk responsibility and acceptance across the enterprise
- A complete risk profile, identifying the full enterprise risk exposure and enabling better utilisation of enterprise resources
- Improved risk awareness throughout the enterprise

Target Audience for This Publication

The intended audience for *COBIT 5 for Risk* is extensive, as are the reasons for adopting and using the framework and the benefits that each enterprise role and function can find in this publication. The roles and functions that are listed in **figure 2** are considered stakeholders for the management of risk. These stakeholders do not necessarily refer to individuals, but to roles and functions within the enterprise or its environment.

Figure 2—*COBIT 5 for Risk* Target Audience and Benefits	
Role/Function	**Benefit of/Reason for Adopting and Adapting *COBIT 5 for Risk***
Board and executive management	• Better understanding of their responsibilities and roles with regard to IT risk management and the implications of IT risk to enterprise strategic objectives • Better understanding of how to optimise IT use for successful strategy execution
Risk function and corporate risk managers for enterprise risk management (ERM)	Assistance with managing IT risk, according to generally accepted ERM principles, and incorporating IT risk into enterprise risk
Operational risk managers	• Linkage of their framework to *COBIT 5 for Risk* • Identification of operational losses or development of key risk indicators (KRIs)
IT management	Better understanding of how to identify and manage IT risk and how to communicate IT risk to business decision makers
IT service managers	Enhancement of their view of operational risk, which should fit into an overall IT risk management framework
Business continuity	Alignment with ERM, because assessment of risk is a key aspect of their responsibility
IT security	Positioning security risk amongst other categories of IT risk
Information security	Positioning IT risk within the enterprise information risk management structure
Chief financial officer (CFO)	Gaining a better view of IT risk and its financial implications for investment and portfolio management purposes
Enterprise governance officers	Assistance with their review and monitoring of governance responsibilities and other IT governance roles
Business	Understanding and management of IT risk—one of many business risk items, all of which should be managed consistently

Figure 2—COBIT 5 for Risk Target Audience and Benefits *(cont.)*	
Role/Function	**Benefit of/Reason for Adopting and Adapting COBIT 5 for Risk**
Internal auditors	Improved analysis of risk in support of audit plans and reports
Compliance	Support with the role as key advisors to the risk function with regard to compliance requirements and their potential impact on the enterprise
General counsel	Support with the role as key advisor for the risk function on regulation-related risk and potential impact or legal implications
Regulators	Support of their assessment of regulated enterprise IT risk management approach and the impact of risk on regulatory requirements
External auditors	Additional guidance on exposure levels when establishing an opinion on the quality of internal control
Insurers	Support with establishing adequate IT insurance coverage and seeking agreement on exposure levels
Rating agencies	In collaboration with insurers, a reference to assess and rate objectively how an enterprise is managing IT risk
IT contractors and subcontractors	• Better alignment of utility and warranty of IT services provided • Understanding of responsibilities arising from risk assessment

Note: The guidance and principles that are provided in this publication are applicable to all enterprises, irrespective of size, industry and nature.

Overview and Guidance on Use of This Publication

COBIT 5 for Risk addresses fundamental questions and issues about IT risk management. **Figure 3** shows these questions and explains how and where *COBIT 5 for Risk* addresses them, if they are within the scope of this guide.

COBIT 5 for Risk refers to the seven enablers of COBIT 5:
• Principles, Policies and Frameworks
• Processes
• Organisational Structures
• Culture, Ethics and Behaviour
• Information
• Services, Infrastructure and Applications
• People, Skills and Competencies

The unique character of each enterprise will result in these enablers being implemented and used in many different ways to manage risk in an optimal manner. This guide provides a pervasive view that explains each concept of COBIT 5 from a risk function perspective through additional guidance and examples.

To facilitate and guide the reader through the comprehensive collection of information, *COBIT 5 for Risk* is divided into three sections and six appendices. Following is a brief description of each section and how those sections are interconnected.

Section 1—Elaborates on risk and risk management and describes briefly how the COBIT 5 principles can be applied to risk management-specific needs. This section provides the reader with a conceptual baseline that is followed throughout the rest of the document.

Section 2—Elaborates on using COBIT 5 enablers for risk management in practice. Governance of enterprise IT (GEIT) is systemic and supported by a set of enablers. In this section, the two perspectives on how to apply the COBIT 5 enablers are explained. Detailed guidance regarding these enablers is provided in the appendices.

Section 2A—Describes the COBIT 5 enablers that are required to build and sustain a risk function.

Section 2B—Describes how the core risk management process of identifying, analysing and responding to risk can be assisted by the COBIT 5 enablers. This section also provides some generic risk scenarios.

Section 3—Introduces the alignment of *COBIT 5 for Risk* with relevant IT or ERM standards and practices, including COSO ERM, ISO 31000, ISO/IEC 27005 and ISO Guide 73. This section also includes a comparison between *COBIT 5 for Risk* and these standards.

Figure 3—*COBIT 5 for Risk* Overview

Question	Where to Find Guidance
What is risk?	**Section 1** defines **risk**, and describes briefly how the COBIT 5 principles can be applied to risk management-specific needs.
How do the COBIT 5 enablers relate to providing risk management?	In general, two perspectives on how to use COBIT 5 in a risk context can be identified: • The **risk function perspective**—Describes what is needed in an enterprise to build and sustain efficient and effective core risk governance and management activities • The **risk management perspective**—Describes how the core risk management process of identifying, analysing and responding to risk, can be assisted by the COBIT 5 enablers These perspectives are introduced in **section 1.3.1**.
How do I set up and maintain an efficient risk function? What is the risk function perspective?	**Section 2A** provides guidance on what is needed to set up and maintain an effective and efficient risk function. It does so by listing and briefly describing the COBIT 5 enablers required, e.g., processes, organisational structures, culture, ethics and behaviour. Putting these enablers in place will result in a performance risk function adding value for the enterprise. **Appendix B** includes detailed descriptions for each enabler listed in section 2A.
How does risk relate to the COBIT 5 principles?	Risk is defined as the potential of enterprise objectives not being achieved or as any opportunity that can enhance enterprise objectives. Maintaining an optimal risk level is one of the three components of the overall value creation objectives of an enterprise, which in turn is supported by the five COBIT 5 principles.
What are key aspects from risk management in practice?	Key components of practical risk management are the risk scenarios. In **section 2B**, risk scenarios and all related topics are explained.
Are there any practical examples of risk scenarios and how to address them?	Yes. **Section 2B**, **chapter 4** contains a comprehensive list of example IT-related risk scenarios. The section also contains some practical advice on how to best use these example scenarios. In **appendix D,** a set of detailed examples is given on how risk scenarios from each category can be mitigated by using a combination of COBIT 5 enablers.
How does *COBIT 5 for Risk* help me in responding to risk?	*COBIT 5 for Risk* makes the link between risk scenarios and an appropriate response. If the response of choice is "mitigate", COBIT 5 contains a wealth of 'controls'—enablers in COBIT 5 terminology—that can be put in place to respond to the risk. **Appendix D** contains a comprehensive set of examples on how that can be done in practice.
Does COBIT 5 align with risk management standards?	Yes. A detailed comparison, in the form of a mapping or qualitative description, is included in **section 3**. The following related standards are discussed in this section: ISO 31000, ISO/EC 27005, COSO ERM and ISO Guide 73.
Does *COBIT 5 for Risk* help me in defining detailed risk analysis methods?	No. Additional guidance on detailed risk analysis methods, taxonomies, tools, etc., is available from multiple sources, including ISACA.

Appendices—Contain the glossary and detailed guidance for the enablers introduced in section 2:
• **Appendix A**—Glossary
• **Appendix B**—Detailed information on enablers for risk governance and management regarding the enablers:
  – B.1—Principles, Policies and Frameworks
  – B.2—Processes
  – B.3—Organisational Structures
  – B.4—Culture, Ethics and Behaviour
  – B.5—Information
  – B.6—Services, Infrastructure Applications
  – B.7—People, Skills and Competencies

- **Appendix C**—Detailed description of core risk management processes
- **Appendix D**—Risk scenarios guidance, containing a comprehensive set of examples on how to mitigate risk scenarios using COBIT 5 enablers
- **Appendix E**—Comparison between *COBIT 5 for Risk* and the legacy *Risk IT Framework*
- **Appendix F**—Template for risk scenario description

Prerequisite Knowledge

COBIT 5 for Risk builds on COBIT 5. Most key concepts of COBIT 5 are repeated and elaborated on, making this guide a fairly standalone book—in essence, not requiring any prerequisite knowledge. However, an understanding of COBIT 5 and its enablers at the foundation level will accelerate the understanding of this guide.

If readers wish to know more about the COBIT 5 concepts beyond what is required for risk management purposes, they are referred to the COBIT 5 framework.

COBIT 5 for Risk also refers to the *COBIT Process Assessment Model (PAM): Using COBIT 5* and the COBIT 5 processes described therein. If readers want to know more about the COBIT 5 processes, e.g., to implement or improve some of them as part of a risk response, they are referred to the *COBIT 5: Enabling Processes* publication.

The COBIT 5 product set also includes a process capability model that is based on the internationally recognised ISO/IEC 15504 Software Engineering—Process Assessment standard. Even though the process assessment model is not prerequisite knowledge for *COBIT 5 for Risk*, readers can use this model as a means to assess the performance of any of the governance or management processes and to identify areas for improvement.

Page intentionally left blank

SECTION 1. RISK AND RISK MANAGEMENT

CHAPTER 1
THE GOVERNANCE OBJECTIVE: VALUE CREATION

Enterprises exist to create value for their stakeholders. Consequently, any enterprise, commercial or not, has value creation as a governance objective.

Value creation means realising benefits at an optimal resource cost **while optimising risk (figure 4)**. Benefits can take many forms, e.g., financial for commercial enterprises or public service for government entities.

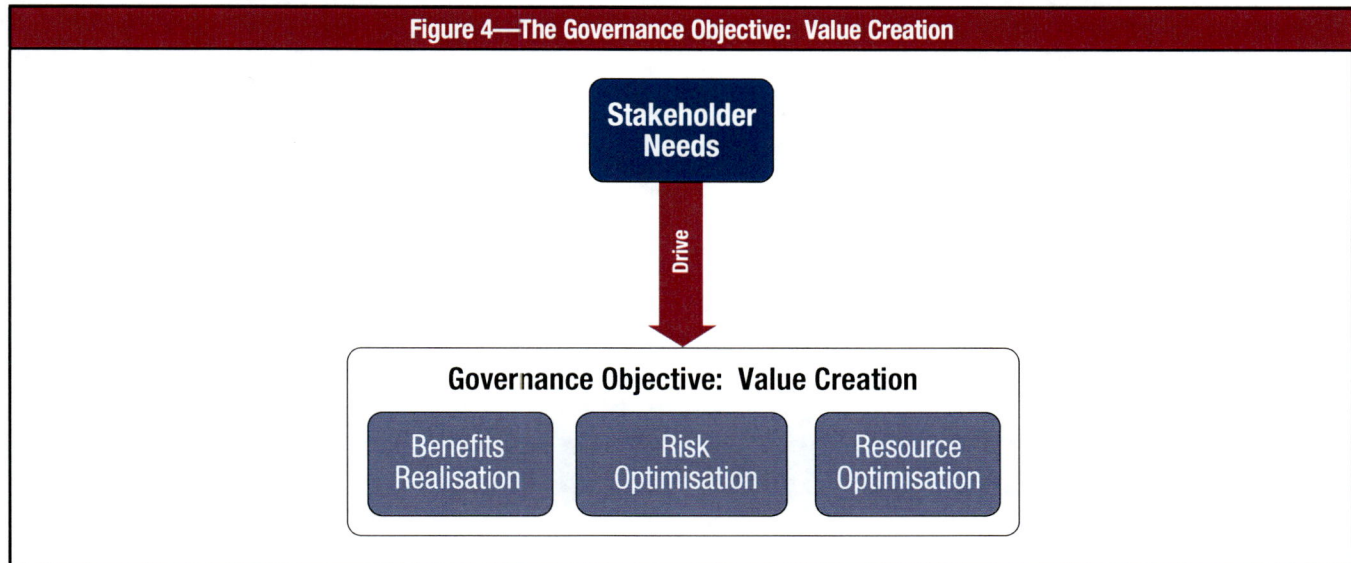

Figure 4—The Governance Objective: Value Creation

Enterprises have many stakeholders, and 'creating value' means different, and sometimes conflicting, things to each stakeholder. Governance is about negotiating and deciding amongst different stakeholder value interests.

The risk optimisation component of value creation shows that:
• Risk optimisation is an essential part of any governance system.
• Risk optimisation cannot be seen in isolation, i.e., actions taken as part of risk management will influence benefits realisation and resource optimisation.

Page intentionally left blank

Page intentionally left blank

CHAPTER 2
RISK

Risk is generally defined as the combination of the probability of an event and its consequence (ISO Guide 73). Consequences are that enterprise objectives are not met. *COBIT 5 for Risk* defines IT risk as business risk, specifically, the business risk associated with the use, ownership, operation, involvement, influence and adoption of IT within an enterprise. IT risk consists of IT-related events that could potentially impact the business. IT risk can occur with both uncertain frequency and impact and creates challenges in meeting strategic goals and objectives.

IT risk always exists, whether or not it is detected or recognised by an enterprise.

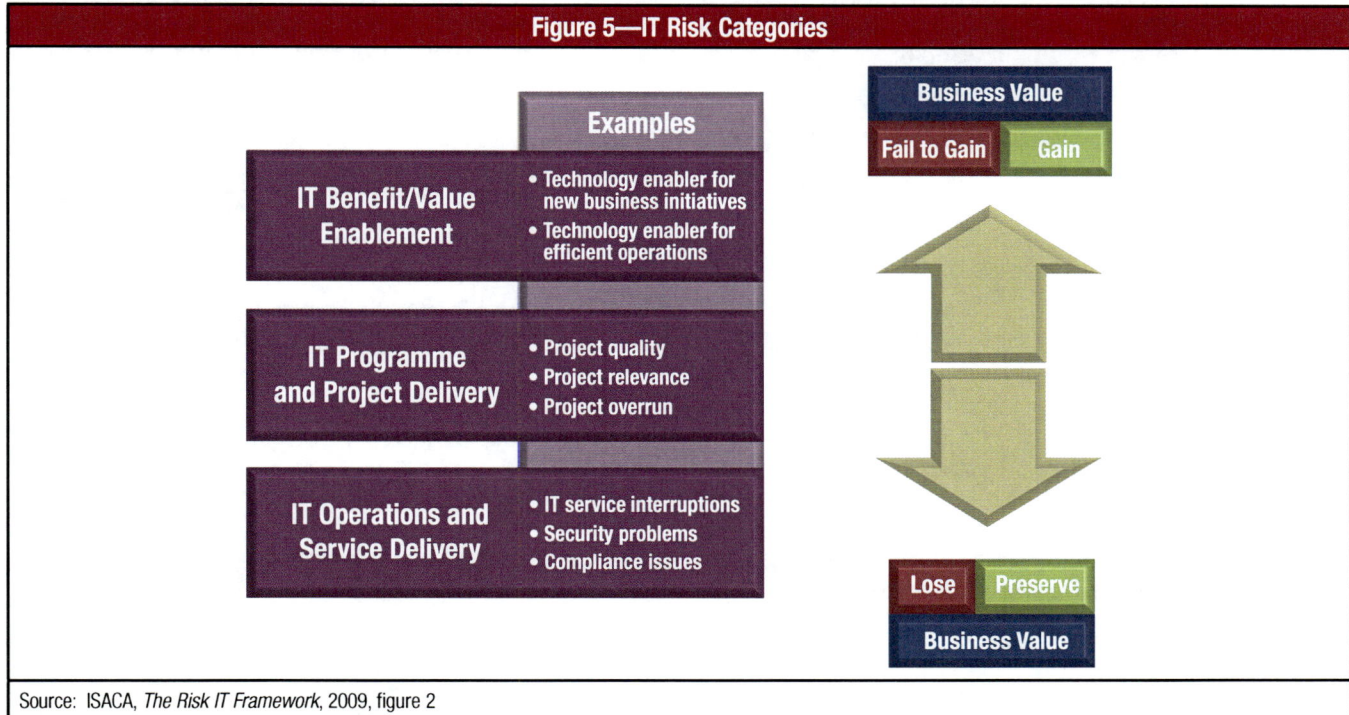

Figure 5—IT Risk Categories

Source: ISACA, *The Risk IT Framework*, 2009, figure 2

IT risk can be categorised as follows:
- **IT benefit/value enablement risk**—Associated with missed opportunities to use technology to improve efficiency or effectiveness of business processes or as an enabler for new business initiatives
- **IT programme and project delivery risk**—Associated with the contribution of IT to new or improved business solutions, usually in the form of projects and programmes as part of investment portfolios
- **IT operations and service delivery risk**—Associated with all aspects of the business as usual performance of IT systems and services, which can bring destruction or reduction of value to the enterprise

Figure 5 shows that for all categories of downside IT risk ('Fail to Gain' and 'Lose' business value) there is an equivalent upside ('Gain' and 'Preserve' business). For example:
- **Service delivery**—If service delivery practices are strengthened, the enterprise can benefit, e.g., by being ready to absorb additional transaction volumes or market share.
- **Project delivery**—Successful project delivery brings new business functionality.

It is important to keep this upside/downside duality of risk in mind (see **figure 6**) during all risk-related decisions. For example, decisions should consider:
- The exposure that may result if a risk is not mitigated versus the benefit if the associated loss exposure is reduced to an acceptable level.
- The potential benefit that may accrue if opportunities are taken versus missed benefits if opportunities are foregone.

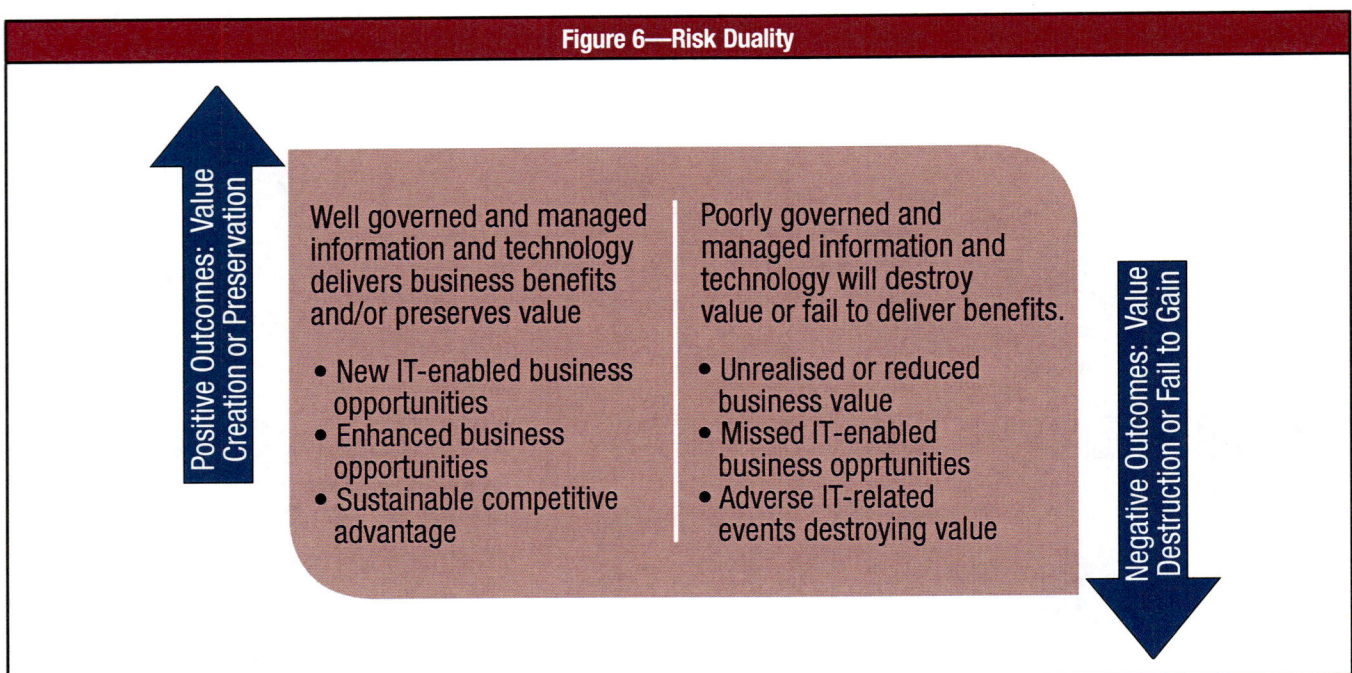

Figure 6—Risk Duality

Positive Outcomes: Value Creation or Preservation

Well governed and managed information and technology delivers business benefits and/or preserves value

- New IT-enabled business opportunities
- Enhanced business opportunities
- Sustainable competitive advantage

Poorly governed and managed information and technology will destroy value or fail to deliver benefits.

- Unrealised or reduced business value
- Missed IT-enabled business opprtunities
- Adverse IT-related events destroying value

Negative Outcomes: Value Destruction or Fail to Gain

Risk is not always to be avoided. Doing business is about taking risk that is consistent with the risk appetite, i.e., many business propositions require IT risk to be taken to achieve the value proposition and realise enterprise goals and objectives, and this risk should be managed but not necessarily avoided.

When risk is referenced in *COBIT 5 for Risk*, it is the **current** risk. The concept of inherent risk is rarely used in *COBIT 5 for Risk*. **Figure 7** shows how inherent, current and residual risk interrelate. Theoretically, *COBIT 5 for Risk* focuses on current risk because, in practice, that is what is used.

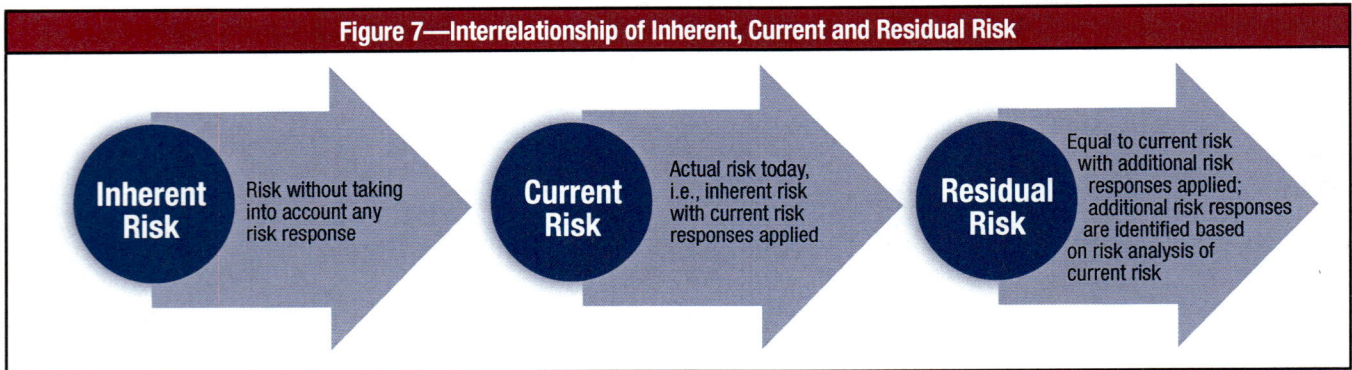

Figure 7—Interrelationship of Inherent, Current and Residual Risk

Inherent Risk — Risk without taking into account any risk response

Current Risk — Actual risk today, i.e., inherent risk with current risk responses applied

Residual Risk — Equal to current risk with additional risk responses applied; additional risk responses are identified based on risk analysis of current risk

CHAPTER 3
SCOPE OF THIS PUBLICATION

3.1 Perspectives on Risk With COBIT 5

The following two perspectives on how to use COBIT 5 in a risk context are shown in **figure 8**:
- **Risk function perspective**—Describes what is needed in an enterprise to build and sustain efficient and effective core risk governance and management activities.
- **Risk management perspective**—Describes how the core risk management process of identifying, analysing, responding to and reporting on risk can be assisted by the COBIT 5 enablers.

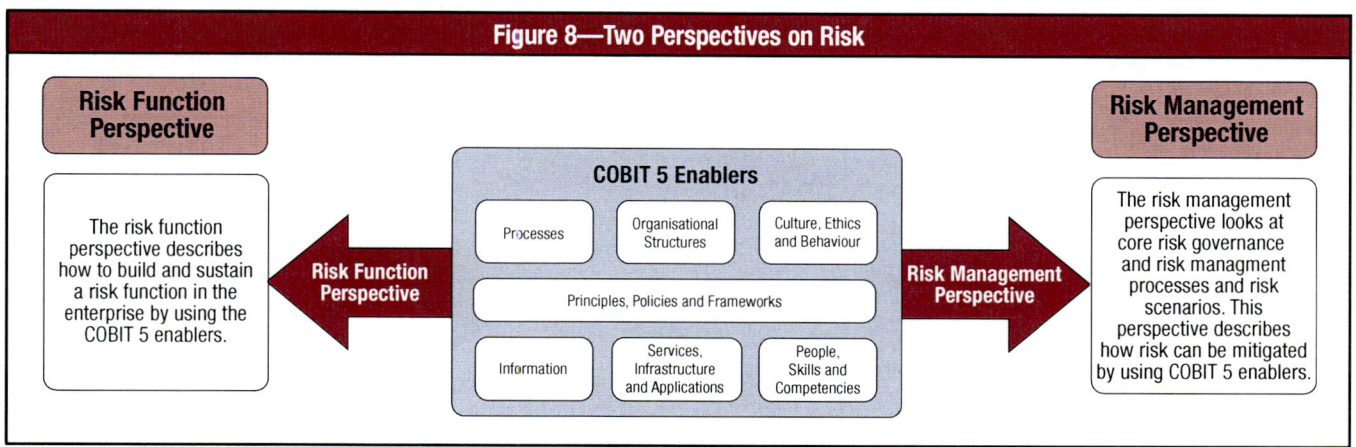

Figure 8—Two Perspectives on Risk

The Risk Function Perspective
COBIT 5 is an end-to-end framework that considers optimisation of risk as a key value objective. COBIT 5 considers governance and management of risk as part of the overall governance and management of enterprise IT.

For each enabler, the risk function perspective describes how the enabler contributes to the overall risk governance and management function. For example, which:
- Processes are required to define and sustain the risk function, govern and manage risk—EDM01, APO01, etc.
- Information flows are required to govern and manage risk—risk universe, risk profile, etc.
- Organisational structures are required to govern and manage risk—ERM committee, risk function, etc.

Section 2A, chapters 2 through 8 contain examples for each enabler. These examples are further elaborated in appendix B.

The Risk Management Perspective
The risk management perspective addresses governance and management, i.e., how to identify, analyse and respond to risk and how to use the COBIT 5 framework for that purpose. This perspective requires core risk processes (COBIT 5 processes EDM03 *Ensure risk optimisation* and APO12 *Manage risk*) to be implemented. These processes are described in detail in appendix C.

Risk is represented by the risk scenarios. A link/comparison is made from the risk management perspective to the COBIT 5 enablers to illustrate how the COBIT 5 framework can help enterprises to govern and manage identified risk.

Illustration of the Two Risk Perspectives

Figure 9 illustrates the two perspectives on risk in an example start-up enterprise, showing the risk setup phases and the risk function perspective and risk management perspective actions for each phase.

Phase	Risk Function Perspective—Using the COBIT 5 enablers to define and sustain an effective risk function	Risk Management Perspective—Using COBIT 5 to identify, analyse, respond to and report on risk
Risk function set-up phase	The enterprise sets up a number of organisational structures, amongst others, a risk function, and assigns risk-related responsibilities to some existing roles, e.g., the chief executive officer (CEO) and the board. The enterprise defines a budget for the risk function, assigns responsibility and accountability to people with the relevant skills, etc.	
Risk management process set-up	The enterprise defines and maintains a risk governance process, i.e., COBIT 5 process EDM03, in the context of a risk management framework, which includes setting risk appetite and tolerance levels, promoting a risk-aware culture, monitoring the risk profile, etc. The enterprise defines and implements a risk management process, i.e., COBIT 5 process APO12. A risk management policy is written.	The enterprise executes the defined processes (EDM03 and APO12), which are supported by the enablers that have been implemented. Based on the processes listed above and the defined risk appetite, the enterprise determines that the quality of their software applications and the security of both software and hardware are key risk issues that require appropriate action. The enterprise responds to risk, i.e., executes the COBIT 5 process practice APO12.06. This response requires the implementation of all previously defined and approved risk response actions. In practice, these risk response actions consist of many of the COBIT 5 enablers, now applied to the overall IT environment.
Risk management in operations		In response to issues with software quality, the enterprise implements/improves the following: • Processes APO09 and APO10 to manage suppliers and service agreements with their suppliers • Process APO11 to manage quality of software development • All processes in the build, acquire and implement (BAI) domain • Process DSS01 to provide IT operations • Process DSS04 to provide business continuity In addition, the other related enablers, e.g., information items, organisational structures, policies, are defined and implemented. In response to issues with security, the enterprise implements/improves the following: • Process APO13 Manage Security • Process DSS05 Manage Security Services In addition, the other related enablers (e.g., information, organisational structures, policies) are defined, implemented and reported on.

Figure 9—Illustration of Two Perspectives on Risk

3.2 Scope of *COBIT 5 for Risk*

Figure 10 shows the scope of *COBIT 5 for Risk* and how it relates to other ISACA documents that—together with *COBIT 5 for Risk*—provide comprehensive guidance on risk governance and management over enterprise IT. **Figure 10** shows that *COBIT 5 for Risk*:
• Focuses on applying the COBIT 5 enablers to risk, through the risk function perspective, i.e., how to use the COBIT 5 enablers to ensure an effective and efficient risk governance and management function.
• Provides high-level guidance on how to identify, analyse and respond to risk through application of the core risk management processes in COBIT 5 and through the use of risk scenarios.

- Aligns with established ERM market reference sources (standards, frameworks and practice guidance) and the ERM initiatives. *COBIT 5 for Risk* includes links/comparisons to the major market reference sources.
- Provides a link between risk scenarios and COBIT 5 enablers that can be used to mitigate risk. (Other IT management frameworks, such as ITIL and ISO/IEC 27001/2, can also be used for that purpose, but no detailed links/mappings are included in this guide.)

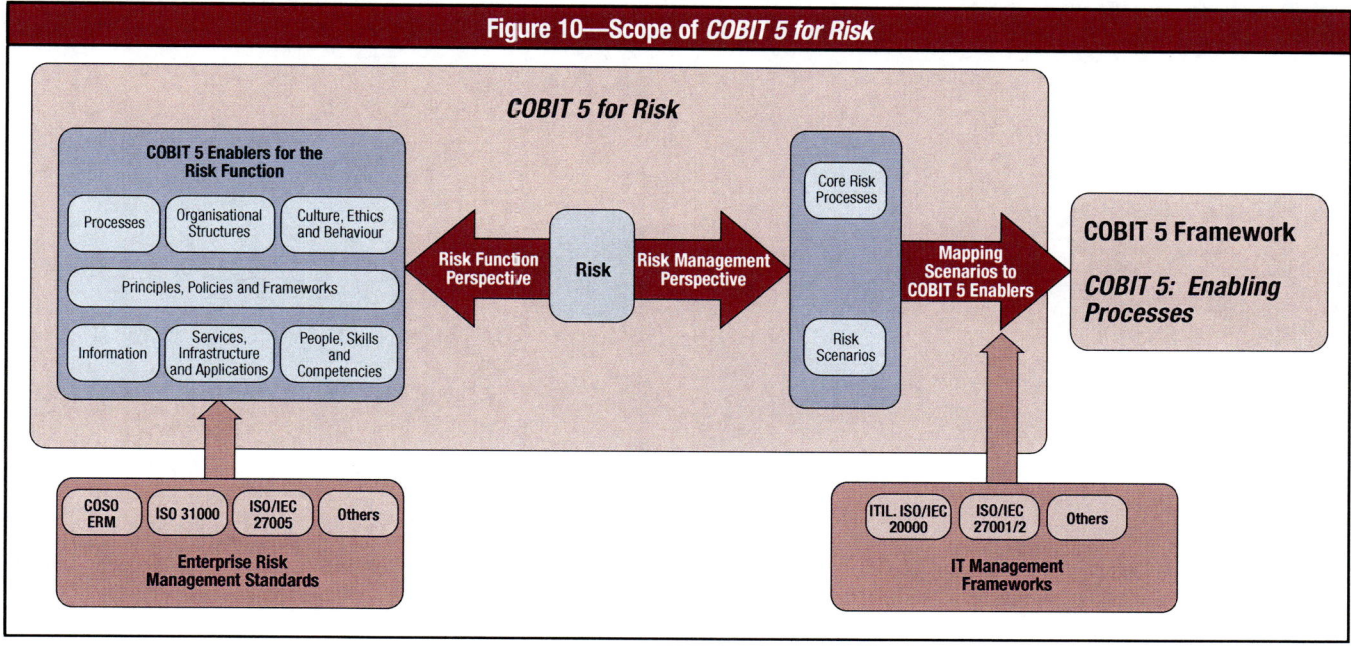

Figure 10—Scope of *COBIT 5 for Risk*

Page intentionally left blank

CHAPTER 4
APPLYING THE COBIT 5 PRINCIPLES TO MANAGING RISK

COBIT 5 is based on five principles, as shown in **figure 11**.

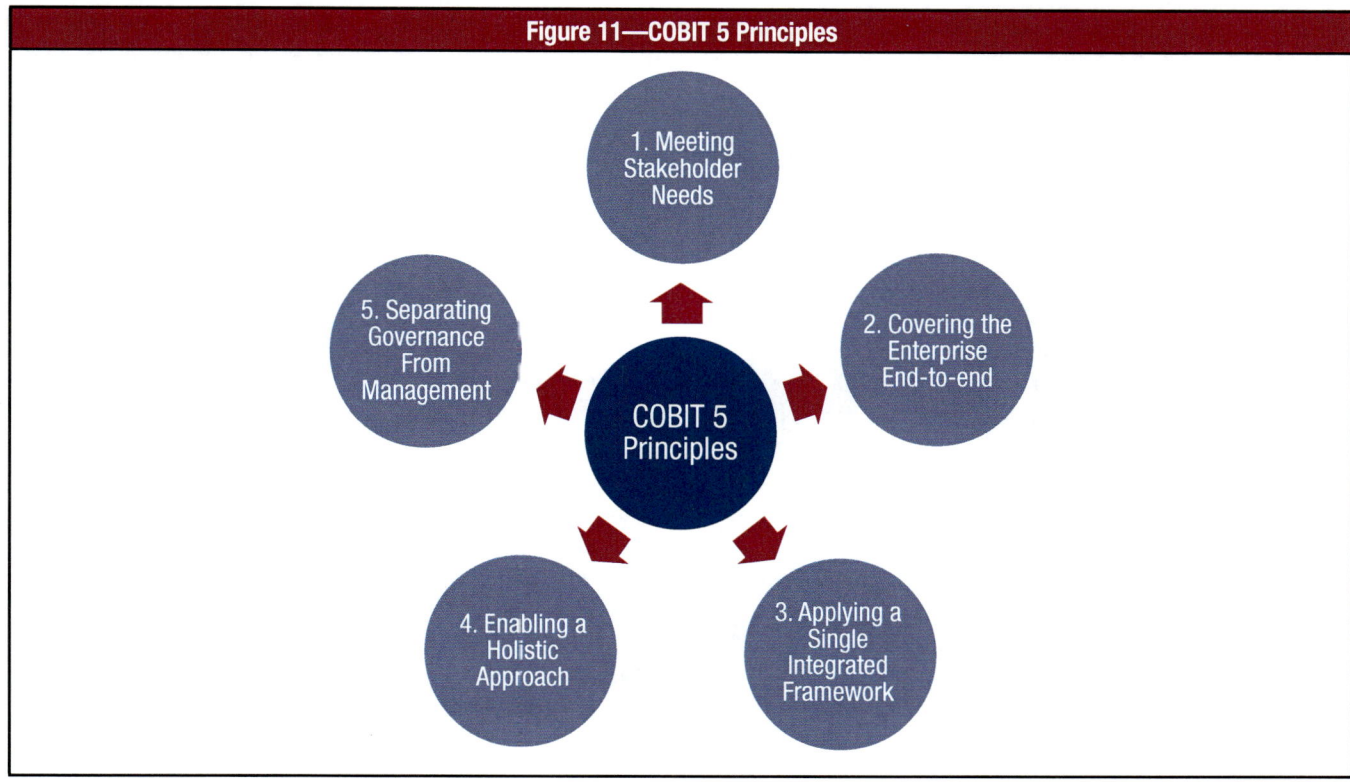

Figure 11—COBIT 5 Principles

COBIT 5 for Risk applies these principles to risk, as follows:
1. **Meeting Stakeholder Needs**—The purpose of risk governance and risk management is to help ensure that enterprise objectives are achieved throughout the goals cascade. Optimising risk is one of the three components of the overall value creation objective for an enterprise.
2. **Covering the Enterprise End-to-end**—*COBIT 5 for Risk* covers all governance and management enablers in its scope and describes all required phases of risk governance and risk management.
3. **Applying a Single Integrated Framework**—*COBIT 5 for Risk* aligns with all major risk management frameworks and standards.
4. **Enabling a Holistic Approach**—*COBIT 5 for Risk* identifies all interconnected elements of the enablers that are required to adequately provide risk governance and management, presenting a holistic and systemic approach towards risk.
5. **Separating Governance from Management**—COBIT 5 distinguishes between risk governance and risk management activities.

4.1 Meeting Stakeholder Needs

Ensuring optimal exposure levels is one of the core objectives of any enterprise, and hence providing adequate risk management concerns a lot of stakeholders, both internal and external to the enterprise. **Figure 2** (see the Executive Summary section Target Audience for This Publication) lists the key stakeholders for risk management and explains their interest in risk management.

4.2 Covering the Enterprise End-to-end

COBIT 5 integrates GEIT into enterprise governance, as follows:
- Covers all functions and processes within the enterprise. COBIT 5 does not focus on only the IT function, but treats information and related technologies as assets that need to be addressed like any other asset, by everyone in the enterprise.
- Considers all IT-related governance and management enablers to be enterprisewide and end-to-end, i.e., inclusive of everyone and everything, internal and external, that are relevant to governance and management of enterprise information and related IT.

To apply this principle to risk, *COBIT 5 for Risk* addresses all enterprise stakeholders, functions and processes that are relevant to risk governance and management.

4.3 Applying a Single Integrated Framework

Many IT-related standards and best practices provide guidance on a subset of IT-related activities. COBIT 5 provides complete enterprise coverage, providing a basis to integrate effectively other frameworks, standards and practices. It serves as a consistent and consolidated source of guidance in a non-technical common language. COBIT 5 aligns with other relevant standards and frameworks, and thus allows the enterprise to use it as the overarching governance and management framework for enterprise IT.

More specifically, *COBIT 5 for Risk* brings together knowledge previously dispersed over other ISACA frameworks (COBIT, Business Model for Information Security [BMIS], Risk IT and Val IT) with guidance from other major risk-related standards, such as ISO 31000, ISO/IEC 27005, and COSO ERM.

The Risk IT Framework was published in 2009 by ISACA with a supporting practitioner guide. The Risk IT risk practices are mapped to governance and management practices in COBIT 5 in appendix A in *COBIT 5: Enabling Processes*. *COBIT 5 for Risk* presents two perspectives on risk, risk function and risk management, to further assist the ERM professional.

4.4 Enabling a Holistic Approach

Efficient and effective governance and management of enterprise IT requires a holistic approach, taking into account several interacting components. COBIT 5 defines a set of enablers to support the implementation of a comprehensive governance and management system for enterprise IT. Enablers are factors that influence, individually and collectively, whether something will work—in this case, governance and management over enterprise IT. Enablers are driven by the COBIT 5 goals cascade, i.e., higher-level enterprise goals and IT-related goals define what the different enablers should achieve. The COBIT 5 framework defines seven categories of enablers (**figure 12**).

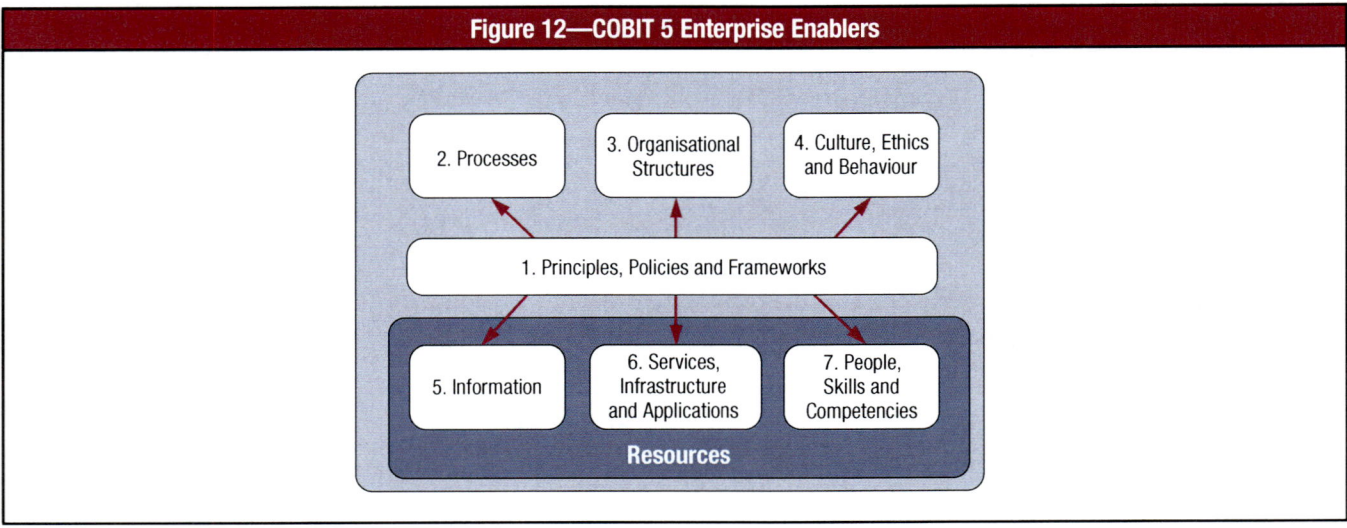

Figure 12—COBIT 5 Enterprise Enablers

The seven categories of enablers also apply to managing risk. The enablers support the provisioning of risk governance and management over enterprise IT, as shown in the following examples:

1. **Principles, Policies and Frameworks**—Risk principles, risk policies and compliance approaches
2. **Processes**—The core risk processes in the evaluate, direct and monitor (EDM) and align, plan and organise (APO) domains, as well as the application of many other processes to the risk function
3. **Organisational Structures**—ERM committee, chief risk officer (CRO)
4. **Culture, Ethics and Behaviour**—Enterprisewide behaviour, management behaviour and risk professionals' behaviour supporting risk management
5. **Information**—Risk profile, risk scenarios, risk map
6. **Services, Infrastructure and Applications**—Emerging risk advisory services
7. **People, Skills and Competencies**—CRISC certification, risk management and technical skills

Section 2A discusses all interconnected enablers required for adequately providing risk governance and management, presenting a holistic and systemic approach towards risk.

4.5 Separating Governance From Management

The COBIT 5 framework makes a clear distinction between governance and management. These two disciplines encompass different types of activities, require different organisational structures and serve different purposes.

Governance

> **Governance ensures that stakeholder needs, conditions and options are evaluated to determine balanced, agreed-on enterprise objectives to be achieved; setting direction through prioritisation and decision making; and monitoring performance and compliance and progress against agreed-on direction and objectives.**

In most enterprises, governance is the responsibility of the board of directors under the leadership of the chairperson.

Good governance means that risk optimisation is part of the governance arrangements that are put in place and risk information is included in the decision-making process. At the same time, the risk function needs to be governed, i.e., provided with direction and monitored. In *COBIT 5 for Risk*, the EDM03 process ensures risk optimisation and is supported by related enablers.

Management

> **Management plans, builds, runs and monitors activities in alignment with the direction set by the governance body to achieve the enterprise objectives.**

In most enterprises, management is the responsibility of the executive management under the leadership of the CEO.

In *COBIT 5 for Risk*, the APO12 process, together with the enablers, allows the enterprise to build, run and monitor an efficient and effective risk management function.

Page intentionally left blank

SECTION 2A. THE RISK FUNCTION PERSPECTIVE

CHAPTER 1
INTRODUCTION TO ENABLERS

1.1. Introduction

This section describes how the COBIT 5 enablers, as introduced in the previous section, can be applied in practical situations and how these enablers can be used to implement effective and efficient risk governance and management in the enterprise.

All enablers defined in COBIT 5 have a set of common dimensions (**figure 13**), which:
• Provide a simple and structured way to deal with enablers
• Allow management of the complex enabler interactions
• Facilitate successful outcomes of the enablers

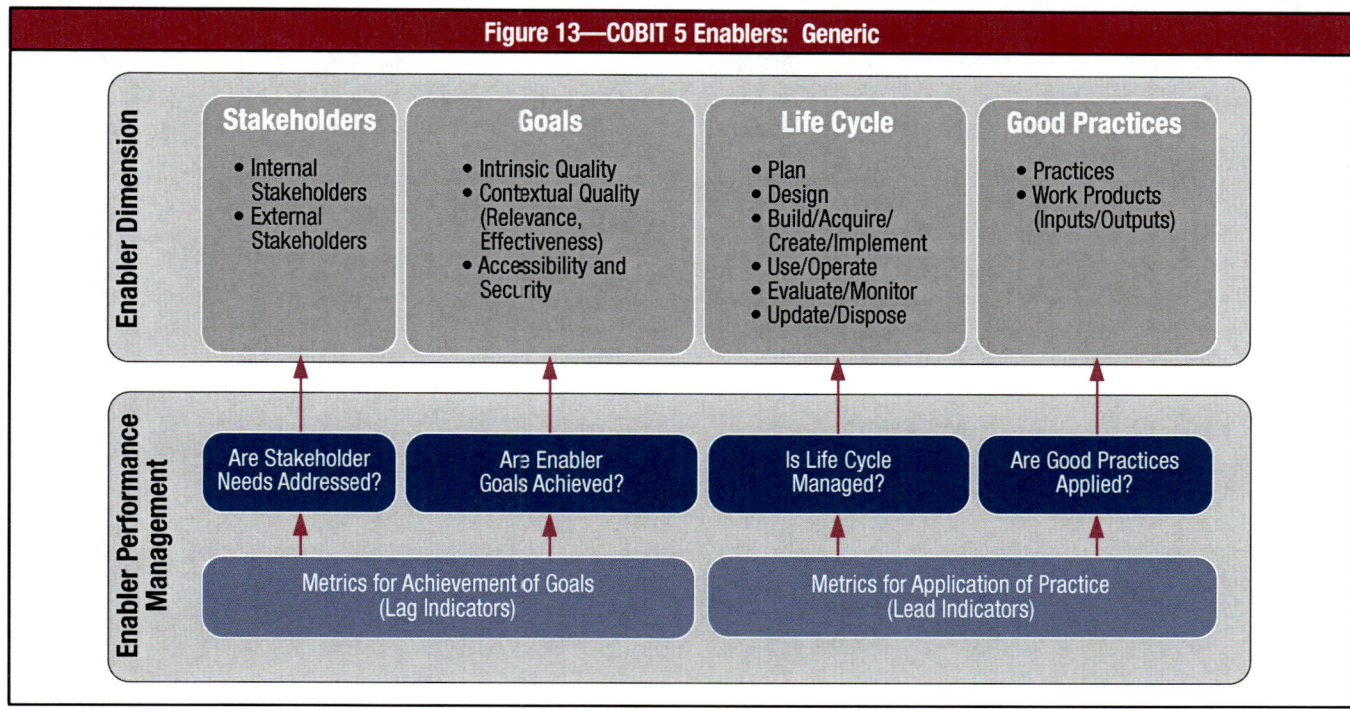

Figure 13—COBIT 5 Enablers: Generic

1.2. Dimensions of the Generic Enabler Model

The four common dimensions for enablers are:
• **Stakeholders**—Each enabler has stakeholders, which are parties who play an active role and/or have an interest in the enabler. For example, for the process enabler, various parties execute process activities and/or have an interest in the process outcomes; and for the organisational structures enabler, stakeholders—each with his/her own roles and interests—are part of the structures. Stakeholders can be internal or external to the enterprise, with their own, sometimes conflicting, interests and needs. Stakeholder needs and interests translate to enterprise goals, which, in turn, translate to IT-related goals for the enterprise.
• **Goals**—Each enabler has goals, which are expected outcomes. Enablers provide value by the achievement of these goals. The enabler goals are the final step in the COBIT 5 goals cascade. Goals can be further split up into the following categories:
  – Intrinsic quality—The extent to which enablers provide accurate, objective and reputable results
  – Contextual quality—The extent to which enabler outcomes fit the purpose, given the context in which the enabler operates. For example, outcomes should be relevant, complete, current, appropriate, consistent, understandable, easy to use and agile.
  – Access and security—The extent to which enablers are accessible—available when, and if, needed,—and secured, i.e., access is restricted to those entitled to and needing it.

- **Life cycle**—Each enabler has a life cycle, from inception, through an operational/useful life and until disposal. Risk identification, assessment, mitigation, monitoring and reporting is part of this life cycle. Following are the life cycle phases:
  - Plan (includes concept development and selection)
  - Design
  - Build/acquire/create/implement
  - Use/operate
  - Evaluate/monitor
  - Update/dispose
- **Good practices**—For each enabler, good practices can be defined. Good practices support the achievement of the enabler goals and provide examples or suggestions for how to best implement the enabler and the required products or inputs and outputs. After the good practices are tuned properly and integrated successfully within the enterprise, they can become, through follow-up with changing business needs and proper monitoring, best practices for the enterprise.

1.3 *COBIT 5 for Risk* and Enablers

COBIT 5 for Risk provides specific guidance related to all enablers:
1. Risk **principles, policies and frameworks**
2. **Processes**, including risk-function-specific details and activities
3. Risk-specific **organisational structures**
4. In terms of **culture, ethics and behaviour**, factors determining the success of risk governance
5. Risk-specific **information** types for enabling risk governance and management within the enterprise
6. With regard to **services, infrastructure and applications**, service capabilities required to provide risk and related functions to an enterprise
7. For the **people, skills and competencies** enabler, skills and competencies specific for risk

The following chapters discuss the seven enablers and their place in risk management. Each chapter begins with a description of the enabler model based on the generic enabler model presented in this chapter; however, specific risk management information is added for each of the enablers as required. Hence, although all of the models resemble each other, they are different and each should be studied carefully.

A risk-specific description of the enabler components and detailed guidance regarding these enablers can be found in appendix B.

CHAPTER 2
ENABLER: PRINCIPLES, POLICIES AND FRAMEWORKS

This chapter contains guidance, from the risk function perspective, on how principles, policies and frameworks can enable risk governance and management in the enterprise. This chapter discusses the following items:
• The Principles, Policies and Frameworks model
• Selected principles, policies and frameworks that are relevant to risk governance and management
• Description of the detailed information that is provided in appendix B.1 for each principle, policy and framework item

2.1 The Principles, Policies and Frameworks Model

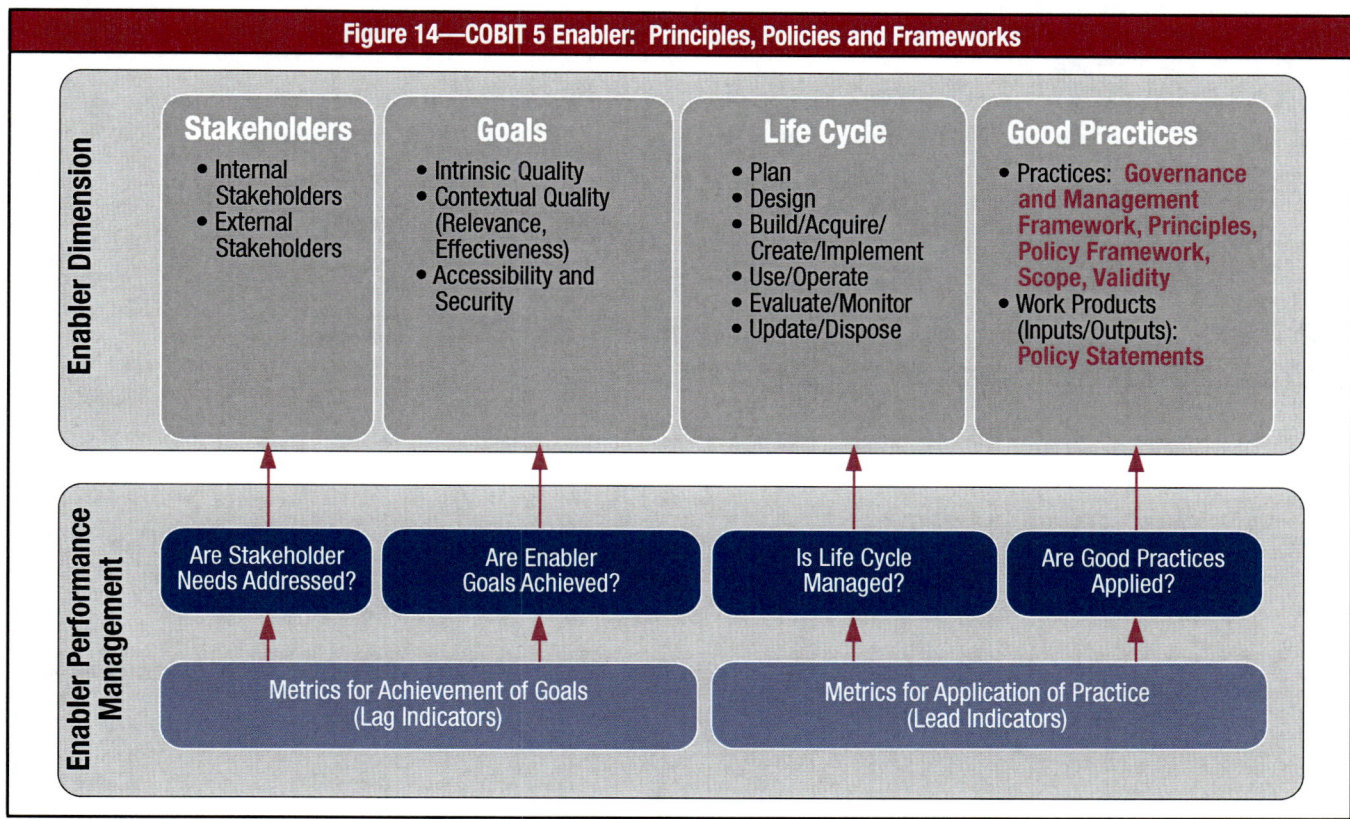

Figure 14—COBIT 5 Enabler: Principles, Policies and Frameworks

The Principles, Policies and Frameworks model (**figure 14**) shows:
• **Stakeholders**—Stakeholders for principles and policies can be internal or external to the enterprise. They include the board and executive management, compliance officers, risk managers, internal and external auditors, service providers and customers, and regulatory agencies. The stakes are twofold: Some stakeholders define and set policies, others have to align to, and comply with, the policies.
• **Goals and metrics**—Principles, policies and frameworks are instruments to communicate the rules of the enterprise, in support of the governance objectives and enterprise values, as defined by the board and executive management. Principles need to:
 – Be limited in number
 – Use simple language, expressing as clearly as possible the core values of the enterprise

Policies provide more detailed guidance on how to put principles into practice and they influence how decision making aligns with the principles. Good policies are:
 – Effective—They achieve the stated purpose.
 – Efficient—They ensure that principles are implemented in the most efficient way.
 – Non-intrusive—They appear logical for those who have to comply with them, i.e., they do not create unnecessary resistance.
 – Aligned—They are in alignment with the overall enterprise strategy.

Access to policies—Is there a mechanism in place that provides easy access to policies for all stakeholders? In other words, do stakeholders know where to find policies?

- **Life cycle**—Policies have a life cycle that has to support the achievement of the defined goals. Frameworks are key because they provide a structure to define consistent guidance. For example, a policy framework provides the structure in which a consistent set of policies can be created and maintained, and it also provides an easy point of navigation within and amongst individual policies.
- **Good practices**:
  - Good practice requires that policies be part of an overall governance and management framework, providing a (hierarchical) structure into which all policies should fit and clearly make the link to the underlying principles.
  - As part of the policy framework, the following items need to be described:
    - Scope and validity
    - Roles and responsibilities of the stakeholders
    - The consequences of failing to comply with the policy
    - The means for handling exceptions
    - The manner in which compliance with the policy will be checked and measured
  - Generally recognised governance and management frameworks can provide valuable guidance on the actual statements to be included in policies.
  - Policies should be aligned with the enterprise's risk appetite. Policies are a key component of an enterprise's system of internal control, whose purpose it is to manage and contain risk. As part of risk governance activities, the enterprise's risk appetite is defined, and this risk appetite should be reflected in the policies. A risk-averse enterprise has stricter policies than a risk-aggressive enterprise.
  - Policies need to be revalidated and/or updated at regular intervals to ensure relevance to business requirements and practices.

2.2 Risk Function Perspective: Principles and Policies Related to Risk Governance and Management

The purpose of this section is to identify and discuss all principles and policies that are required to build and sustain effective and efficient IT risk governance and IT risk management in an enterprise.

The risk principles (**figure 15**) focus on providing a systematic, timely and structured approach to risk management, which will contribute to consistent, comparable and reliable results. The risk principles formalise and standardise the implementation of the risk policies, which are discussed in section 2.2.2. Detailed descriptions of the risk principles are in appendix B.1.

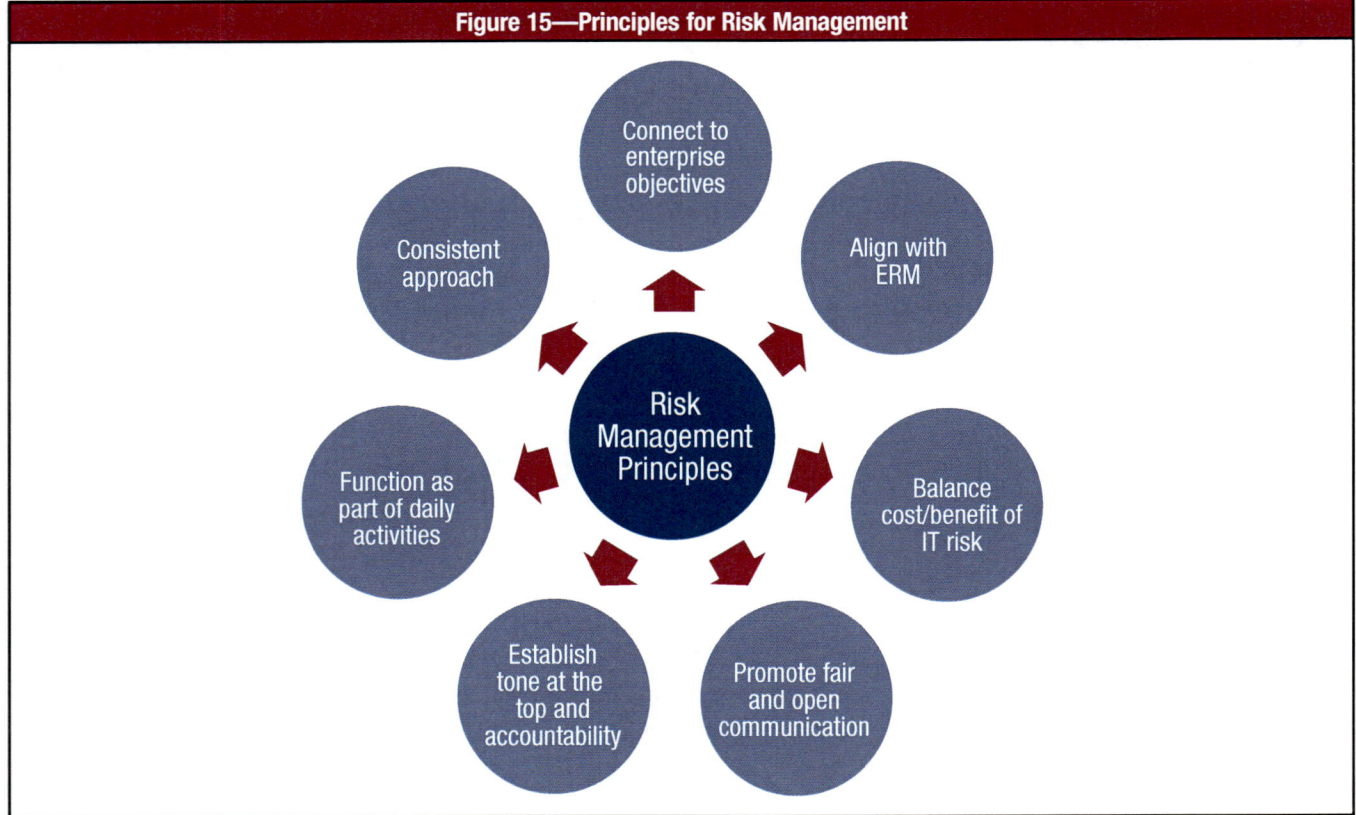

Figure 15—Principles for Risk Management

Policies provide more detailed guidance on how to put principles into practice and how they will influence decision making. Not all relevant policies are written and owned by the IT risk function. A number of risk-related policies are described in this publication, and the policy driver within the enterprise is specified.

The core IT risk policy is shown in **figure 16**, which also provides examples of operational-level policies that should be considered for complete IT risk management. Depending on the size and nature of the enterprise, these may be policies in their own right, or sections within existing policies. For operational efficiency, it is necessary to keep these policies in sync with the risk policy.

Figure 16—Risk Policy Examples	
Policies	**Description**
Core IT risk policy	Defines, at strategic, tactical and operational levels, how the risk of an enterprise needs to be governed and managed pursuant to its business objectives. This policy translates enterprise governance into risk governance principles and policy and elaborates risk management activities.
Information security policy	Sets behavioural guidelines in protecting corporate information and the associated systems and infrastructure. The business requirements regarding security and storage are more dynamic than IT risk management, so, for effectiveness, their governance needs to be handled separately from the governance of IT risk. However, for operational efficiency, it is necessary to keep the information security policy in sync with the IT risk policy.
Crisis management policy	As with IT security, network management and data security, IT crisis management is one of the operational level policies that needs to be considered for complete IT risk management. It sets the guidelines on how to act in situations of crisis and details the sequence in which to deal with each of the identified (key) areas of risk.
Third-party IT service delivery management policy	Sets guidelines for managing the risk related to third-party services. It sets out a framework of expectations in behaviour and security precautions taken by third-party service providers to manage the risk related to the service provision.
Business continuity policy	Contains management commitment and view on: • Business impact analysis (BIA) • Business contingency plans with trusted recovery • Recovery requirements for critical systems • Defined thresholds and triggers for contingencies • How to handle escalation of incidents • Disaster recovery plan (DRP) • Training and testing
Programme/project management policy	Deals with managing risk linked to projects and programmes. It details management position and expectation regarding programme and project management. Moreover, it handles accountability, goals and objectives regarding performance, budget, risk analysis, reporting and mitigating adverse events during the execution of programmes and projects.
Human resources (HR) policies	Detail what employees can expect from the enterprise and what the enterprise expects from employees. They provide detailed acceptable and unacceptable behaviour by employees, and, in doing so, manage the risk that is linked to human behaviour.
Fraud risk policy	Is concerned with protecting the enterprise brand, reputation and assets from loss and/or damage, resulting from incidents of fraud and/or misconduct. The policy provides guidance to all employees on reporting any suspicious activity and ways to handle sensitive information and evidence. It helps to raise an anti-fraud culture and raise awareness of the risk.
Compliance policy	Explains the assessment process regarding compliance with regulatory, contractual and internal requirements. It lists roles and responsibilities for the different activities in the process and provides guidance on metrics to be used to measure compliance.
Ethics policy	Defines the essentials of how people within an enterprise will interact with one another, as well as how they will interact with any customers or clients they serve.
Quality management policy	Details management vision on the quality objectives of the enterprise, the acceptable level of quality and the duties of specific departments to ensure quality.
Service management policy	Provides direction and guidance to ensure the effective management and implementation of all information technology services to meet the business and customer requirements, within a framework of performance measurement. It also deals with management of risk related to IT services. Detailed guidance on service management and the optimisation of risk related to services is included in the ITIL V3 framework.
Change management policy	Communicates management intent that changes to the enterprise information technology be managed and implemented in a way that minimizes risk and impact to the stakeholders. The policy contains information on the assets in the scope and the established standard change management process.

Figure 16—Risk Policy Examples *(cont.)*	
Policies	**Description**
Delegation of authority policy	Details: • The authority that the board strictly retains for itself • The general principles of delegation of authority • A schedule of the delegation of authority (including clear boundaries) • A clear definition of the organisational structures to which the board delegates its authority
Whistle-blower policy	Should: • Encourage employees to raise concerns and questions • Provide avenues for employees to raise concerns in full confidence • Ensure employees will receive a response to raised concerns and be able to escalate a concern if they are not satisfied with the response • Reassure the employees that they are protected when they raise issues and should not be afraid of reprisals
Internal control policy	The purpose is to: • Communicate management internal control objectives • Establish standards for the design and operation of the enterprise system of internal controls to reduce the exposure to all risk faced by the enterprise
Intellectual property (IP) policy	The purpose is to ensure that all risk related to the use, ownership, sale and distribution of the outputs of IT-related creative endeavours by employees of an enterprise, e.g., software development, is detailed in an appropriate way, from the start of any endeavour.
Data privacy policy	A statement or a document that discloses the ways that a party gathers, uses, discloses and manages personal data. Personal information can be anything that can be used to identify an individual, including but not limited to name, address, date of birth, marital status, contact information, ID issue and expiry date, financial records, credit information, medical history, travel destination, and intentions to acquire goods and services. The policy defines how an enterprise collects, stores, and releases the personal information that it collects. The policy informs the client of the specific information that is collected and whether it is kept confidential, shared with partners, or sold to other firms or enterprises. Furthermore, the policy ensures compliance with relevant legislation related to data protection.

CHAPTER 3
ENABLER: PROCESSES

This chapter contains guidance on how processes can enable risk governance and management in the enterprise (the risk function perspective). To that purpose, the following items are discussed:
• The Processes model
• A list of all processes relevant for the governance and management of risk. This list is a subset of the COBIT 5 processes as described in *COBIT 5: Enabling Processes*, and it is grouped in two categories of processes:
  – Key supporting processes for risk governance and management
  – Other supporting processes for risk governance and management
  Note: These processes do not contain the core risk management processes (EDM03 and APO12, i.e., describing risk analysis, risk response). They are described in section 2B of this guide and in appendix C.)
• A description of the detailed information that is provided for each process. The actual detailed information is included in appendix B.2.

3.1 The Processes Model

A process is defined as 'a collection of practices influenced by the enterprise's policies and procedures that takes inputs from a number of sources (including other processes), manipulates the inputs and produces outputs (e.g., products, services)'.

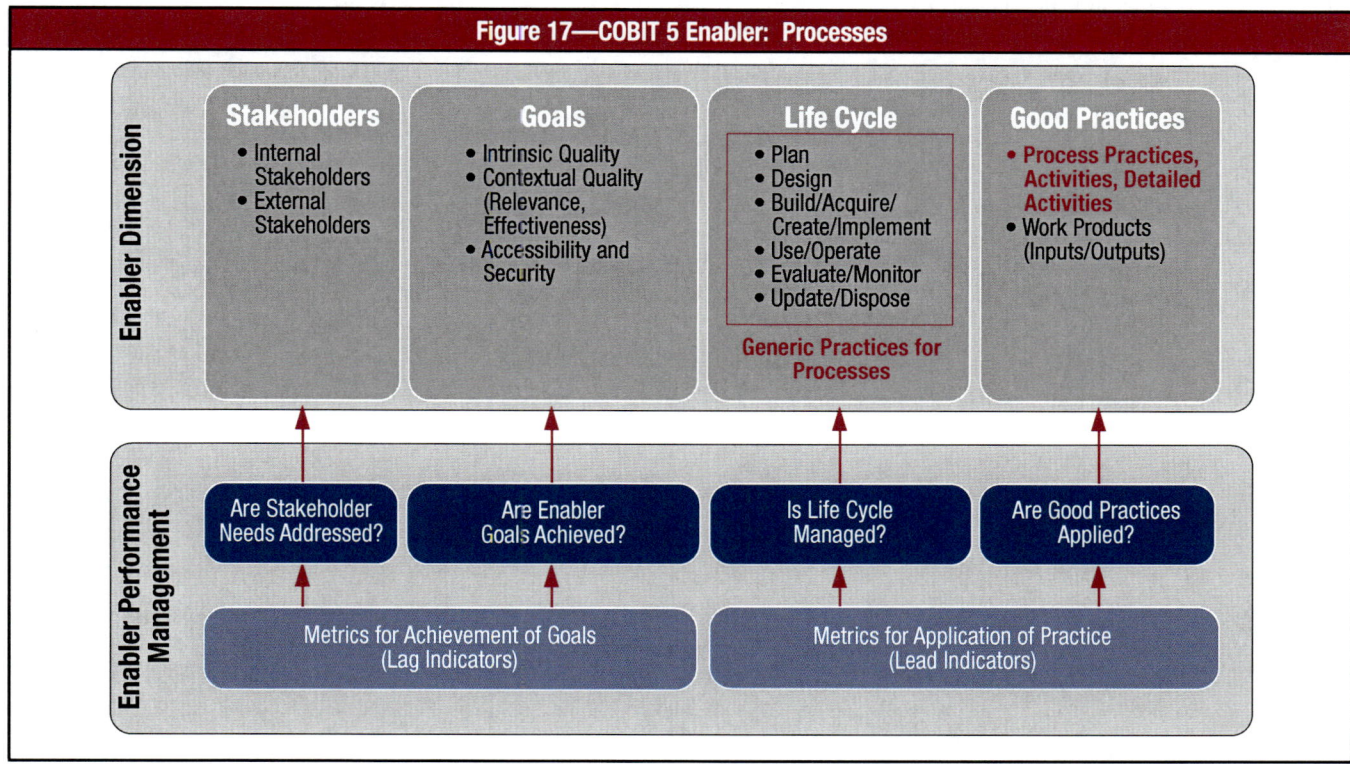

Figure 17—COBIT 5 Enabler: Processes

The Processes model (**figure 17**) shows:
- **Stakeholders**—Processes have internal and external stakeholders, with their own roles; stakeholders and their responsibility levels are documented in RACI (responsible, accountable, consulted, informed) charts. External stakeholders include customers, business partners, shareholders and regulators. Internal stakeholders include the board, management, staff and volunteers.
- **Goals**—Process goals are defined as 'a statement describing the desired outcome of a process. An outcome can be an artefact, a significant change of a state or a significant capability improvement of other processes'. They are part of the goals cascade, i.e., process goals support IT-related goals, which, in turn, support enterprise goals. Process goals can be categorised as:
  - Intrinsic goals—Does the process have intrinsic quality? Is it accurate and in line with good practice? Is it compliant with internal and external rules?
  - Contextual goals—Is the process customised and adapted to the enterprise's specific situation? Is the process relevant, understandable and easy to apply?
  - Accessibility and security goals—The process remains confidential, when required, and is known and accessible to those who need it.
- **Life cycle**—Each process has a life cycle. It is defined, created, operated, monitored and adjusted/updated or retired. Generic process practices such as those defined in the COBIT process assessment model (PAM), based on International Organization for Standardization/International Electrotechnical Commission (ISO/IEC) 15504, can assist with defining, running, monitoring and optimising processes.
- **Good practices**—*COBIT 5: Enabling Processes* contains a process reference model, in which process internal good practices are described in increasing levels of detail: practices, activities and detailed activities. In this publication, this good practice is not repeated; only risk-specific guidance is developed when relevant.

3.2 Risk Function Perspective: Processes Supporting the Risk Function

The purpose of this section is to identify and discuss all COBIT 5 processes that are required to build and sustain effective and efficient risk governance and risk management in an enterprise. In other words, it lists all processes that will support the risk function.

These processes do not include the core processes (EDM03 and APO12) to govern and manage risk. These are described in detail in section 2B of this guide.

The detailed process-related and risk-specific information for the COBIT 5 processes includes:
- **Process goals and metrics**—For each process, a limited number of risk-specific process goals are included, and for each process goal, a limited number of risk-specific example metrics are listed, reflecting the clear relationship between the goals and the metrics.
- **Detailed description of the process practices**—This description contains, for each practice:
  - Risk-specific practice inputs and outputs (work products), with indication of origin and destination
  - Risk-specific process activities

Figure 18 highlights the key supporting COBIT 5 processes (shown in dark pink), as well as the other supporting processes (shown in light pink). The paragraphs following **figure 18** provide a short description of each supporting process, the reason it is important and the key outputs.

Note: The core risk processes (shown in light blue) are detailed in section 2B, chapter 1.

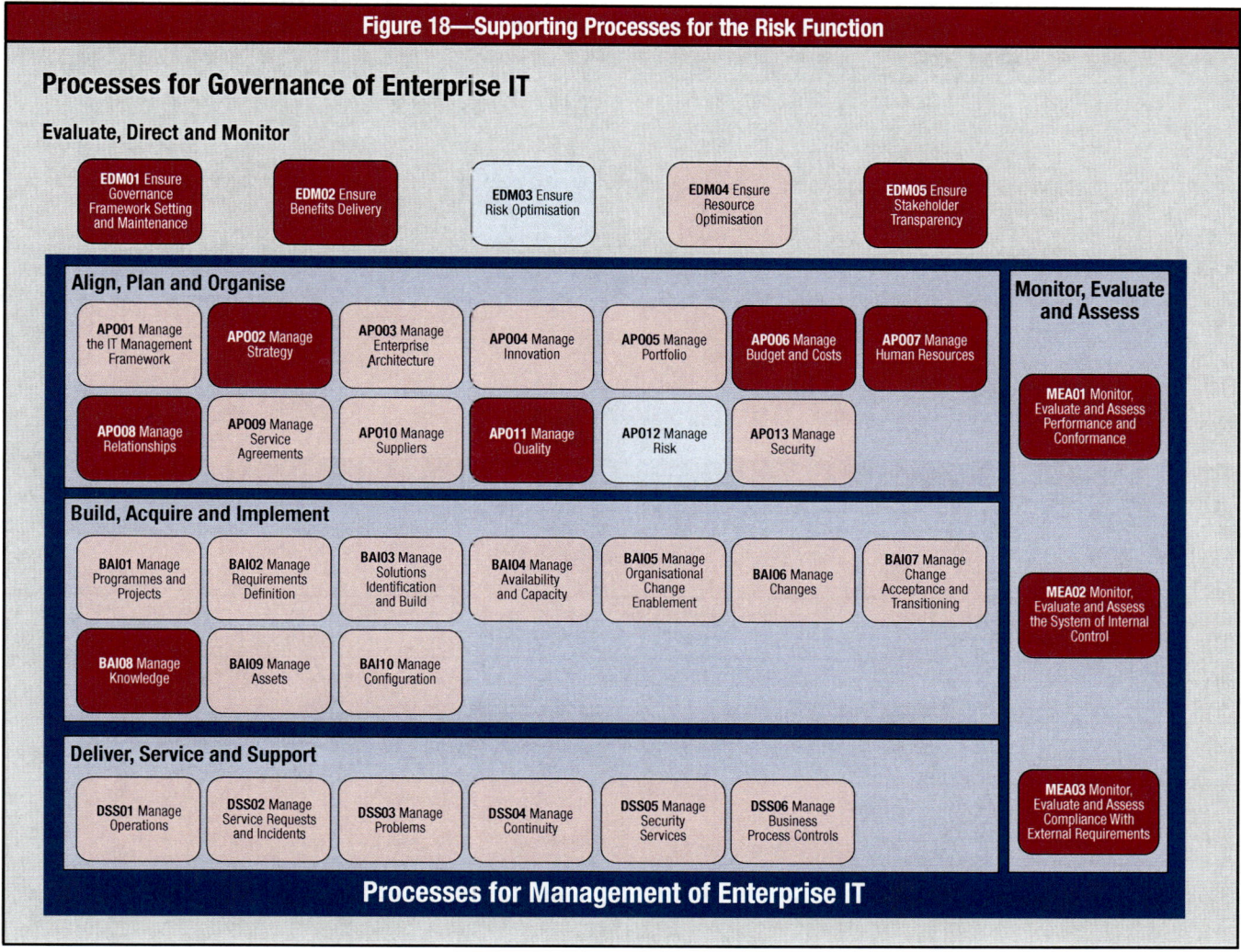

Figure 18—Supporting Processes for the Risk Function

The processes listed in **figure 19** are key supporting processes for the risk function in the enterprise.

Figure 19—Key Supporting Processes for the Risk Function		
Process Identification	**Reasoning**	**Risk-specific Outputs**
EDM01 Ensure Governance Framework Setting and Maintenance	Governing and managing risk requires the setup of an adequate governance framework to put in place enabling structures, principles, processes and practices.	Risk governance guiding principles
EDM02 Ensure Benefits Delivery	This process focuses on managing the value that the risk function generates.	Actions to improve risk value delivery
EDM05 Ensure Stakeholder Transparency	The enterprise risk function requires transparent performance and conformance measurement, with goals and metrics approved by stakeholders.	Evaluation of risk reporting requirements
APO02 Manage Strategy	IT risk management strategy must be well defined and aligned to the ERM approach.	Risk management strategy
APO06 Manage Budget and Costs	The risk function needs to be budgeted.	Financial and budgetary requirements
APO07 Manage Human Resources	Risk management requires the right amount of people, skills and experience.	HR competencies framework
APO08 Manage Relationships	Maintain the relationships between the risk function and the business.	Risk management communication plan
APO11 Manage Quality	Quality is a not to be a disregarded component of effective risk management. Risk deliverables should be treated following the enterprise quality management system	Quality review of risk deliverables
BAI08 Manage Knowledge	The risk function needs to be provided with the knowledge required to support staff in their work activities.	• Classification of risk function information • Access control over information • Rules for disposal of information

Figure 19—Key Supporting Process for the Risk Function *(cont.)*		
Process Identification	**Reasoning**	**Risk-specific Outputs**
MEA01 Monitor, Evaluate and Assess Performance and Conformance	Risk is a key aspect in the monitoring, evaluating and assessing of business and IT.	Risk monitoring metrics and targets
MEA02 Monitor, Evaluate and Assess the System of Internal Control	Internal controls are key in monitoring and containing risk to avoid risk becoming an issue.	Results of internal control monitoring and reviews
MEA03 Monitor, Evaluate and Assess Compliance with External Requirements	Compliance with laws, regulations and contractual requirements represent risk and have to be monitored, evaluated and assessed in alignment with enterprise strategy.	Reports of non-compliance issues and root causes

The other supporting processes, listed in **figure 20**, are not described in any further detail in this guide. The standard COBIT 5 process guidance is included in *COBIT 5: Enabling Processes*.

Figure 20–Other Supporting Processes for the Risk Function	
Process Identification	**Reasoning**
EDM04 Ensure Resource Optimisation	The risk function needs to optimise its resource utilisation.
APO01 Manage the IT Management Framework	The risk management function is supporting the IT management framework.
APO03 Manage Enterprise Architecture	The risk function should use the enterprise architecture as a key source of information to support risk assessments.
APO04 Manage Innovation	The risk function should always be looking for new methodologies, technologies and tools that can support the governance and management of risk within the enterprise.
APO05 Manage Portfolio	The risk portfolio of systems needs to be managed and considered as a main source of information.
APO09 Manage Service Agreements	The risk function can use (internal or external) service providers, e.g., a co-sourced IT risk function.
APO10 Manage Suppliers	The risk function can use (internal or external) service providers, e.g., a co-sourced IT risk function.
APO13 Manage Security	The risk function has security requirements that need management.
BAI01 Manage Programmes and Projects	New risk management software will need to be implemented.
BAI02 Manage Requirements Definition	Requirements for new risk management software will need to be developed.
BAI03 Manage Solutions Identification and Build	New risk management software will need solutions identification and building.
BAI04 Manage Availability and Capacity	Risk management software availability and capacity will need to be managed.
BAI05 Manage Organisational Change Enablement	New risk management software will need change management.
BAI06 Manage Changes	Risk management software will need a defined change process.
BAI07 Manage Change Acceptance and Transitioning	Risk management software will need a user acceptance process defined.
BAI09 Manage Assets	The risk function needs to be involved in the management of its IT assets.
BAI10 Manage Configuration	The risk function needs to manage its IT configuration together with the IT department.
DSS01 Manage Operations	The risk function is supported by IT tools and applications in its daily operations and needs to manage this.
DSS02 Manage Service Requests and Incidents	The risk function needs to follow up on service requests and incidents for its own IT assets.
DSS03 Manage Problems	The risk function needs to follow up on problems regarding its own IT assets.
DSS04 Manage Continuity	The risk function must manage business continuity for itself.
DSS05 Manage Security Services	The risk function needs to comply with security policies regarding its own IT assets.
DSS06 Manage Business Process Controls	The risk function needs to manage business process controls over its IT assets.

CHAPTER 4
ENABLER: ORGANISATIONAL STRUCTURES

This chapter contains guidance on how organisational structures can enable risk governance and risk management in the enterprise (the risk function perspective). To that purpose, the following items are discussed:
• The Organisational Structures model
• A list of all organisational structures relevant to risk governance and risk management. This list is an extended subset of the COBIT 5 lists of roles in *COBIT 5: Enabling Processes*, and it is grouped into two categories of structures, as follows:
  – Core risk governance- and management-related organisational structures
  – Supporting organisational structures
• A description of the detailed information that is provided for each organisational structure. The actual detailed information is included in appendix B.3.

4.1 The Organisational Structures Model

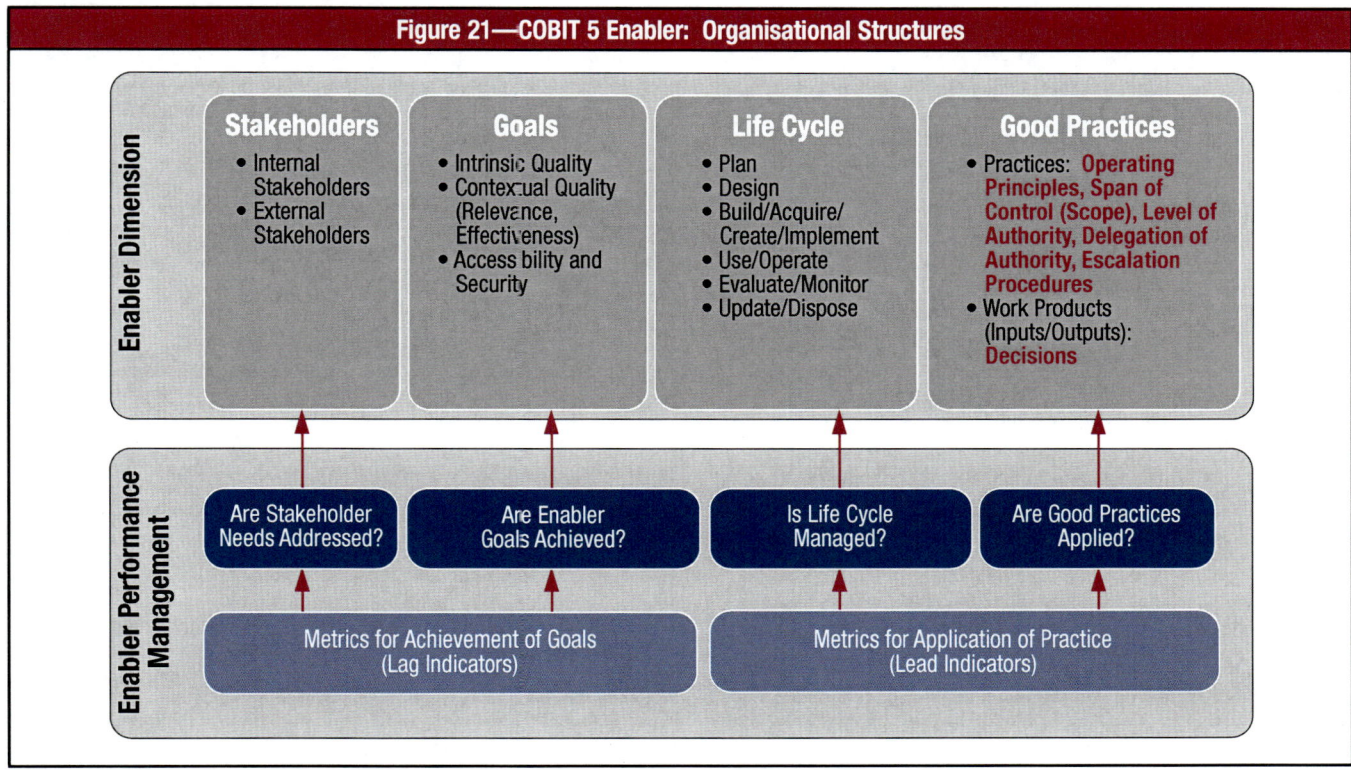

Figure 21—COBIT 5 Enabler: Organisational Structures

The Organisational Structures model (**figure 21**) shows:
• **Stakeholders**—Organisational structures stakeholders can be internal or external to the enterprise. These stakeholders include the individual members of the structure, other structures, organisational entities, clients, suppliers and regulators. Their roles vary and include decision making, influencing and advising. The stakes of each of the stakeholders also vary, i.e., what interest do they have in the decisions made by the structure?
• **Goals**—The goals for the Organisational Structures enabler include have a proper mandate, have well-defined operating principles and the application of other good practices. The outcome of the organisational structures enabler should include numerous good activities and decisions.
• **Life cycle**—An organisational structure has a life cycle. The organisational structure is created, exists and is adjusted, and, finally, it can be disbanded. During its inception, a mandate—a reason and purpose for its existence—has to be defined.
• **Good practices**—A number of good practices for organisational structures can be distinguished, such as:
  – Operating principles—The practical arrangements regarding how the structure operates, such as frequency of meetings, documentation and housekeeping rules
  – Decisions—Risk-based direction considering the processing of relevant inputs and required or expected outputs

- Span of control—The decision-rights boundaries of the organisational structure
- Level of authority/decision rights—The decisions that the structure is authorised to make
- Delegation of authority—The structure's authority to delegate a subset of its decision rights to other structures that report to it.
- Escalation procedures—The escalation path for a structure, which describes the required actions in case of problems in making decisions.

4.2 Risk Function Perspective: Risk Governance- and Management-related Organisational Structures

The purpose of this section is to identify and discuss all organisational structures/roles that are required to build and sustain effective and efficient risk governance and risk management in an enterprise. **Core** structures/roles are those that spend most of their time working on risk governance and management. The supporting roles have the support for risk governance and management as only a small(er) part of their overall job.

Figure 22 defines core organisational structures that support the risk management function.

Figure 22—Key Organisational Structures	
Role/Structure	**Definition/Description**
Enterprise risk management (ERM) committee	The group of enterprise executives that is accountable for the enterprise-level collaboration and consensus required to support ERM activities and decisions. This committee is considered to be the second line of defence against risk manifestation. An IT risk council may be established to consider IT risk in more detail and advise the ERM committee. Committee members are usually drawn from the board and the CEO chairs the committee.
Enterprise risk group	The enterprise risk group considers risk in more detail and advises the ERM committee. The enterprise risk group is a collection of business and IT resources that serve as the risk management programme facilitators and maintain the risk register and risk profile for the enterprise. They are considered the first line of defence against risk manifestation.
Risk function	The most senior official of the enterprise who is accountable for all aspects of risk management across the enterprise, including taking direction from the ERM committee. An IT risk officer function may be established to oversee risk.
Audit department	The enterprise function responsible for provision of internal audit reports on the risk associated with gaps in controls identified while performing reviews.[1] As this is considered the third and last line of defence, a representative can be invited to the ERM committee.
Compliance department	The enterprise function responsible for insight into the enterprise risk related to regulations, legal mandates and internal policies and standards.

The organisational structures in **figure 22** can also be related to the 'three lines of defence' model, as shown in **figure 23**.

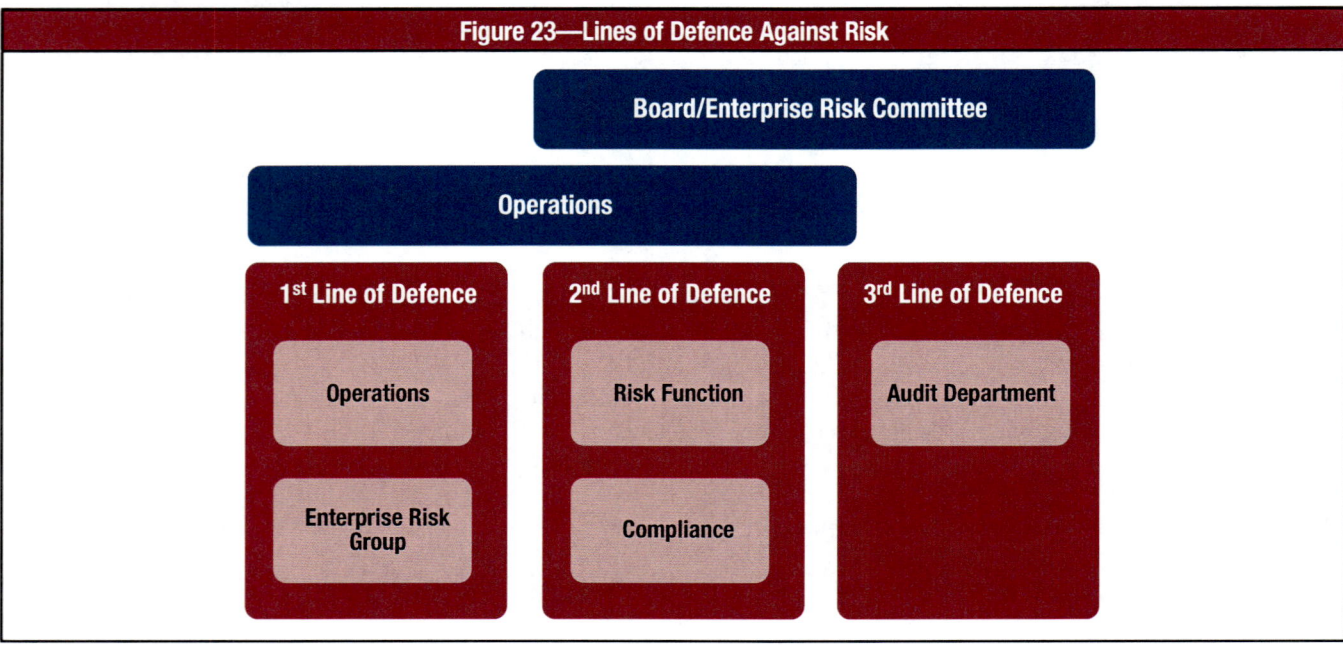

Figure 23—Lines of Defence Against Risk

---

[1] For a detailed description of the 'audit department', refer to appendix B.3.

In the three lines of defence model,[2] management control is the first line of defence in risk management. The various risk control and compliance oversight functions established by management are the second line of defence, and independent assurance is the third line of defence. Each of these three lines of defence plays a distinct role within the enterprise's wider governance framework:

- As the first line of defence, operational managers own and manage risk. They also are responsible for implementing corrective actions to address process and control deficiencies.
- In practice, a single line of defence often can prove inadequate. Management establishes various risk management and compliance functions to help build and/or monitor the first line of defence controls.
- Internal auditors provide the governing body and senior management with comprehensive assurance based on the highest level of independence and objectivity within the enterprise. This high level of independence is not available in the second line of defence. Internal audit also provides assurance on the manner in which the first and second lines of defence achieve risk management.

Although neither governing bodies nor senior management are considered to be among the three lines of defence in this model, they are the primary stakeholders served by the lines of defence. They are the parties best positioned to help ensure that the three lines of defence model is reflected in the enterprise's risk management and control processes.

Senior management and governing bodies collectively have responsibility and accountability for setting the enterprise's objectives, defining strategies to achieve those objectives and establishing governance structures and processes to best manage the risk in accomplishing those objectives.

Additional roles/structures throughout the enterprise can be found to have an effect on risk management. These roles/structures are not directly involved in risk management, but are relevant stakeholders in this process. **Figure 24** lists other relevant roles and structures, and provides a brief description of each and the stake each holds in supporting the risk function.

Figure 24—Other Relevant Structures for Risk		
Role/Structure	**Definition/Description**	**Role in the Risk Process**
Board	The group of the most senior executives and/or non-executives of the enterprise who are accountable for the governance of the enterprise and have overall control of its resources	• Provides oversight on the impact of enterprise risk on the enterprise objectives[3] and makes risk decisions to protect stakeholder value at an optimal resource cost • Sets the tone at the top with regard to risk management and awareness
CEO	The highest-ranking officer who is in charge of the total management of the enterprise	Provides executive support and collaborates on risk management decisions
Strategy (IT executive) committee	A group of senior executives appointed by the board to ensure that the board is involved in and kept informed of major IT-related matters and decisions. The committee is accountable for managing the portfolios of IT-enabled investments, IT services and IT assets, ensuring that value is delivered and risk is managed. The committee is normally chaired by a board member, not by the chief information officer (CIO).	Being proactive in managing risk related to portfolio of IT investments
Chief operating officer (COO)	The most senior official of the enterprise who is accountable for the operations of the enterprise.	Consults on operations risk
Business executive	A senior management individual accountable for the operation of a specific business unit or subsidiary. This includes key business line owners and heads of departments such sales, marketing, human resources, manufacturing, etc.	• Consults on risk related to business lines or departments • Accepts ownership of assigned risk and reports on mitigation progress
CIO/Chief technology officer (CTO)	The most senior official of the enterprise who is responsible for aligning IT and business strategies and accountable for planning, resourcing and managing the delivery of IT services and solutions to support enterprise objectives	Consults about the technical aspects of risk and actions

---

[2] Source: The Institute of Internal Auditors (IIA); *IIA Position Paper: The Three Lines of Defense in Effective Risk Management and Control,* USA, 2013
[3] For more information on role of the board in overseeing risk, refer to 'Risk Governance: Balancing Risk and Rewards', Blue Ribbon Commission report, National Association of Corporate Directors (NACD), 2009.

Figure 24—Other Relevant Structures for Risk *(cont.)*		
Role/Structure	**Definition/Description**	**Role in the Risk Process**
Business process owner	An individual accountable for the performance of a process in realising its objectives, driving process improvement and approving process changes. In general, a business process owner must be at an appropriately high level in the enterprise and have authority to commit resources to process-specific risk management activities.	• Consults on risk related to business processes • Accepts ownership of assigned risk and reports on mitigation progress
Chief information security officer (CISO)	The most senior officer of the enterprise who is accountable for the security of enterprise information in all forms	• Consults on security risk • Co-ordinates incident response
CFO	The most senior official of the enterprise who is accountable for all aspects of financial management, including financial risk and controls and reliable and accurate accounts	Consults on acceptable loss exposure levels, importance of risk factors, and on the cost of risk response options
Business continuity manager	An individual who manages, designs, and/or assesses an enterprise business continuity capability to ensure that the enterprise critical functions continue to operate following disruptive events	Consults on disruption of enterprise operations
IT process (service) owners	The lead individual who is accountable and responsible for the IT process and/or service provided	• Consults on risk related to IT processes or services. • Accepts ownership of assigned risk and reports on mitigation progress.
Service manager	An individual who manages the development, implementation, evaluation and ongoing management of new and existing products and services for a specific customer, user or group of customers	Consults on risk-related service management
Head of HR	The most senior official of an enterprise who is accountable for planning and policies with respect to all human resources in that enterprise.	Consults about human aspects of risk and actions
Privacy officer	An individual who is responsible for monitoring the risk and business impacts of privacy laws and for guiding and co-ordinating the implementation of policies and activities that will ensure that the privacy directives are met	Consults about key aspects related to monitoring of the risk and business impacts of privacy laws and policies
Project and programme steering committee	A group of stakeholders and experts who are accountable for guidance of programmes and projects, including management and monitoring of plans, allocation of resources, delivery of benefits and value and management of programme and project risk	Articulates the risk associated with programmes and projects
Chief insurance officer	The most senior official of the enterprise who is in charge of managing insurance policies	Consulted for risk sharing/transfer
Procurement	The function that engages and manages third parties and associated processes, such as contracts, and the overall supply chain	• Consults and advises in managing third-party risk • Negotiates contract terms to manage risk and performance management

CHAPTER 5
ENABLER: CULTURE, ETHICS AND BEHAVIOUR

This chapter contains guidance on how culture and behaviour can enable risk governance and management in the enterprise (the risk function perspective). To that purpose, the following items are discussed:
• The Culture, Ethics and Behaviour model
• A list of selected behaviours relevant to the provisioning of risk governance and management
• A description of the detailed information that is provided for each behaviour item. The detailed information is included in appendix B.4.

5.1 The Culture, Ethics and Behaviour Model

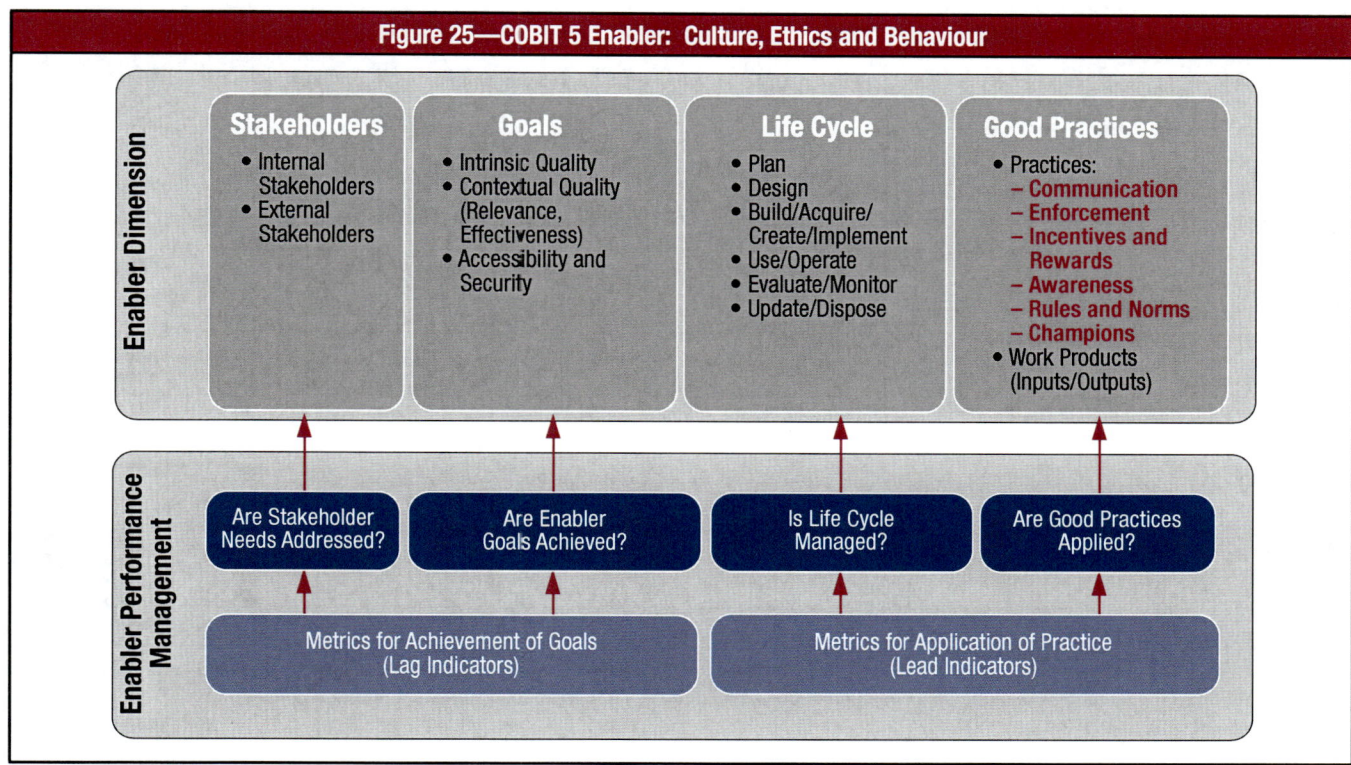

Figure 25—COBIT 5 Enabler: Culture, Ethics and Behaviour

The Culture, Ethics and Behaviour model (**figure 25**) shows:
• **Stakeholders**—Culture, ethics and behaviour stakeholders can be internal or external to the enterprise. Internal stakeholders include the entire enterprise, external stakeholders include regulators, e.g., external auditors or supervisory bodies. Stakes are twofold: Some stakeholders, e.g., legal officers, risk managers, HR managers, remuneration boards and officers, deal with defining, implementing and enforcing desired behaviours; others have to align with the defined rules and norms.
• **Goals**—Goals for the Culture, Ethics and Behaviour enabler relate to:
  – Organisational ethics, determined by the values by which the enterprise wants to live
  – Individual ethics, determined by the personal values of each individual in the enterprise and depending, to an important extent, on external factors such as religion, ethnicity, socioeconomic background, geography and personal experiences
  – Individual behaviours, which collectively determine the culture of an enterprise. Many factors, such as the external factors mentioned above, interpersonal relationships in enterprises, personal objectives and ambitions also drive behaviours. Some types of behaviours that can be relevant in this context include:
    · Behaviour towards taking risk—How much risk does the enterprise feel it can absorb and which risk is it willing to take?
    · Behaviour towards following policy—To what extent will people embrace and/or comply with policy?
    · Behaviour towards negative outcomes—How does the enterprise deal with negative outcomes, i.e., loss events or missed opportunities? Will it learn from them and try to adjust, or will blame be assigned without treating the root cause?

- **Life cycle**—An organisational culture, ethical stance and individual behaviours, etc., all have their life cycles. Starting from an existing culture, an enterprise can identify required changes and work towards their implementation. Several tools, described in the good practices, can be used.
- **Good practices**—Good practices for creating, encouraging and maintaining desired behaviour throughout the enterprise include:
  - Communication throughout the enterprise of desired behaviours and the underlying corporate values
  - Awareness of desired behaviour, strengthened by the example behaviour exercised by senior management and other champions
  - Incentives encourage desired behaviour, and deterrents discourage undesirable behaviour, often as part of the HR reward and recognition programme
  - Re-evaluation of expectations, influences and changes in behaviour and practices reports on existing behaviour versus the behaviour that management perceives
  - Rules and norms, which provide more guidance on desired organisational behaviour. These link very clearly to the principles and policies that an enterprise puts in place.

5.2 Risk Function Perspective: Risk Governance- and Management-related Culture and Behaviour

The purpose of this section is to identify relevant behaviours and culture elements that are required to build and sustain effective and efficient risk management in an enterprise and that contribute to establishing and maintaining a risk-aware culture at all levels of the enterprise. In other words, it lists relevant behaviours that support the risk function.

The desirable behaviours are categorized according to three levels within the enterprise:
- General (enterprisewide)
- Risk professionals
- Management

The goal or key objective, suitable criteria or the desired outcomes are listed for each behaviour (**figure 26**).

Figure 26—Relevant Behaviour for Risk Governance and Management	
Behaviour	**Key Objective/Suitable Criteria/Outcome**
General Behaviour	
Has a risk- and compliance-aware culture throughout, including the proactive identification and escalation of risk.	Must define a risk management approach and risk appetite. Zero tolerance of non-compliance with legal and regulatory requirements must be established.
Has defined policies that have been communicated and that drive behaviour.	All personnel understand and implement the requirements of the enterprise as defined in policies.
Shows positive behaviour towards raising issues or negative outcomes.	Whistle-blowers are seen as having a positive contribution to the enterprise. The 'blame culture' is avoided. Personnel understand the need for risk awareness and reporting possible weaknesses.
Recognises the value of risk.	Personnel understand the importance to the enterprise of maintaining risk awareness and the value that managing risk adds to their role.
Has transparent and participative culture as an important focus point.	Communication is open and overt so that facts are not omitted, misrepresented or understated. The negative impact of hidden agendas is avoided.
Shows mutual respect.	Stakeholders and risk assessors are encouraged to collaborate. People are respected as professionals and treated as experts in their roles.
Business accepts ownership of risk.	Risk practices are incorporated throughout the enterprise. Accountabilities are cleared and accepted. IT-related business risk is owned by the business and not viewed solely as the responsibility of the IT department or the risk function.
Allows risk acceptance as a valid option.	Management understands the likelihood and consequence of the impact of risk acceptance. The impact is determined to be within the enterprise's risk appetite.
Risk Professional Behaviour	
Show effort to understand what risk is for each stakeholder and how it impacts their objectives.	Risk professionals understand the commercial reality of the impact of risk. This may include competitive, operational, regulatory and compliance requirements. Although there may be risk common to a certain industry, each enterprise is unique in terms of how these risk items impact specific enterprise objectives.
Create awareness and understanding of the risk policy.	Alignment between risk capacity, risk appetite and enterprise policy can lead to effective risk strategies.
Collaboration and two-way communication during risk assessment.	Risk assessment is fundamentally accurate, complete and addresses stakeholder needs.

Figure 26—Relevant Behaviour for Risk Governance and Management *(cont.)*	
Behaviour	**Key Objective/Suitable Criteria/Outcome**
Risk Professional Behaviour *(cont.)*	
Risk appetite is clear and communicated in a timely fashion with relevant stakeholders.	Stakeholders manage risk more effectively and there is appropriate alignment with organisational strategy and objectives.
Policies reflect risk appetite and risk tolerance.	Employees and management operate within risk tolerance. Business lines apply formal risk appetite and tolerance to daily practices. There is a clear process for proposing and making changes to risk appetite levels, with senior management consideration and approval.
Enterprise culture supports effective risk practice.	Stakeholders understand risk from common portfolio views (product, process) and apply risk-based decision making to daily practice.
KRIs are used as an early warning.	KRIs are associated with valid metrics and can be used as an indicator of process or control failure. KRI metrics are available and accessible for regular reporting and relate to objectives.
Risk indicators or events that fall outside of appetite and tolerance are acted on.	Risk indicators are linked to the management risk response and remediation activities.
Management Behaviour	
Senior management sets direction and demonstrates visible and genuine support for risk practices.	Quality risk management practices are maintained through genuine support from senior management.
Management engages with all relevant stakeholders to agree on actions and follow up on action plans.	The correct stakeholders are appropriately involved in ensuring timely resolution of issues and the achievement of business plans.
Genuine commitment is obtained and resources are assigned for execution of actions.	Personnel are empowered in executing actions required by risk management decisions.
Management aligns policies and actions to risk appetite.	Management makes appropriate risk decisions in complying with policies. Risk adjusted revenue is in line with management expectations.
Management proactively monitors risk and action plan progress.	Remediation plans are completed within expected business time frames and have a positive impact on enterprise objectives.
Risk trends are reported to management.	The timely reporting of risk trends proactively manages risk and avoids lost opportunities.
Effective risk management is rewarded.	Good risk practice is acknowledged. Employees' performance goals and reward structures are set to stimulate effective risk management practices and appropriate execution of mitigation actions.

In appendix B.4, a detailed description can be found of how culture aspects such as communication, rules, incentives, rewards and raising awareness can influence these behaviours.

Page intentionally left blank

CHAPTER 6
ENABLER: INFORMATION

This chapter contains guidance on how information items can enable risk governance and management in the enterprise (the risk function perspective). To that purpose, the following items are discussed:
• The Information model
• A list of selected information items relevant to risk governance and management
• A description of the detailed information that is provided for each information item. The actual detailed information is included in appendix B.5.

6.1 The Information Model

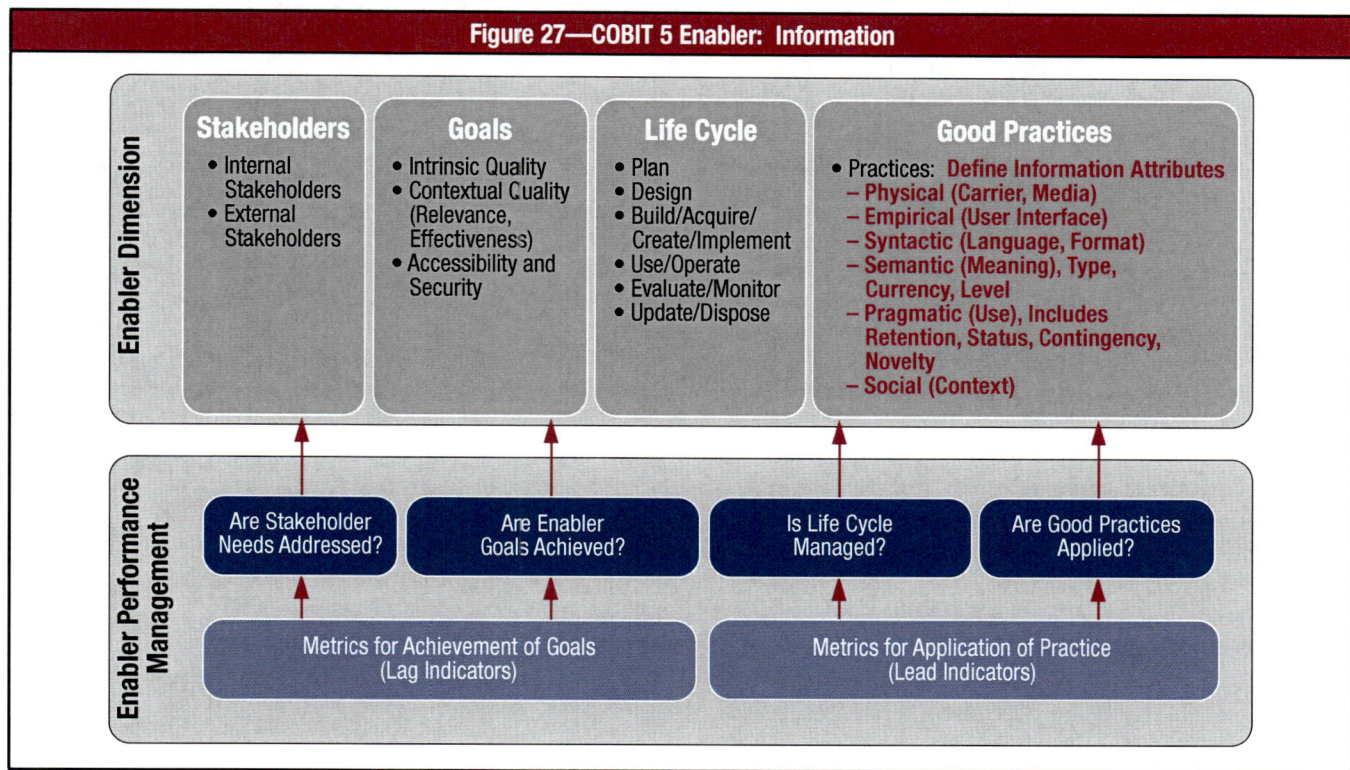

Figure 27—COBIT 5 Enabler: Information

The Information model in **figure 27** shows:
• **Stakeholders**—Can be internal or external to the enterprise. The generic model also suggests that, apart from identifying the stakeholders, their stakes (i.e., why they care or are interested in the information) need to be identified. Different categories of roles in dealing with information are possible, ranging from detailed proposals (e.g., suggesting specific data or information roles such as architect, owner, steward, trustee, supplier, beneficiary, modeller, quality manager, security manager) to more general proposals—for instance, distinguishing amongst information producers, information custodians and information consumers, as follows:
 – Information producer is responsible for creating the information
 – Information custodian is responsible for storing and maintaining the information
 – Information consumer is responsible for using the information

These categories refer to specific activities with regard to the information resource. Activities depend on the life cycle phase of the information; therefore, to find a category of roles that has an appropriate level of granularity for the Information model, the information life cycle dimension of the Information model can be used. This means that information stakeholder roles, e.g., information planners, information obtainers, information users, can be defined in terms of information life cycle phases. At the same time, this means that the information stakeholder dimension is not an independent dimension; different life cycle phases have different stakeholders.

Whereas the relevant roles depend on the information life cycle phase, the stakes can be related to information goals.

- **Goals**—The goals of information are divided into **three subdimensions of quality**:
  - Intrinsic quality—The extent to which data values are in conformance with the actual or true values. It includes:
    - Accuracy—The extent to which information is correct and reliable
    - Objectivity—The extent to which information is unbiased, unprejudiced and impartial
    - Believability—The extent to which information is regarded as true and credible
    - Reputation—The extent to which information is highly regarded in terms of its source or content
  - Contextual and representational quality—The extent to which information is applicable to the task of the information user and is presented in an intelligible and clear manner, recognising that information quality depends on the context of use. It includes:
    - Relevancy—The extent to which information is applicable and helpful for the task at hand
    - Completeness—The extent to which information is not missing and is of sufficient depth and breadth for the task at hand
    - Currency—The extent to which information is sufficiently up to date for the task at hand
    - Appropriate amount of information—The extent to which the volume of information is appropriate for the task at hand
    - Concise representation—The extent to which information is compactly represented
    - Consistent representation—The extent to which information is presented in the same format
    - Interpretability—The extent to which information is in appropriate languages, symbols and units, with clear definitions
    - Understandability—The extent to which information is easily comprehended
    - Ease of manipulation—The extent to which information is easy to manipulate and apply to different tasks
  - Security/accessibility quality—The extent to which information is available or obtainable. It includes:
    - Availability/timeliness—The extent to which information is available when required, or easily and quickly retrievable
    - Restricted access—The extent to which access to information is restricted appropriately to authorised parties
- **Life cycle**—The full life cycle of information needs to be considered, and different approaches may be required for information in different phases of the life cycle. The COBIT 5 Information enabler distinguishes the following phases:
  - Plan—The phase in which the creation and use of the information resource is prepared. Activities in this phase may refer to the identification of objectives, the planning of the information architecture, and the development of standards and definitions, e.g., data definitions, data collection procedures.
  - Design—Defining the risk management information requirements
  - Build/acquire/create/implement—The phase in which the information resource is acquired. Activities in this phase may refer to the creation of data records, the purchase of data and the loading of external files.
  - Use/operate, which includes:
    - Store—The phase in which information is held electronically or in hard copy (or even just in human memory). Activities in this phase may refer to the storage of information in electronic form, e.g., electronic files, databases, data warehouses, or as hard copy, e.g., paper documents.
    - Share—The phase in which information is made available for use through a distribution method. Activities in this phase may refer to the processes involved in getting the information to places where it can be accessed and used, e.g., distributing documents by email. For electronically held information, this life cycle phase may largely overlap with the store phase, e.g., sharing information through database access, file/document servers.
    - Use—The phase in which information is used to accomplish goals. Activities in this phase may refer to all kinds of information usage (e.g., managerial decision making, running automated processes), and may also include activities such as information retrieval and converting information from one form to another.

Information is an enabler for enterprise governance; hence, information use as defined in the Information model can be thought of as the purposes for which enterprise governance stakeholders need information when assuming their roles, fulfilling their activities and interacting with each other.

The interactions between stakeholders require information flows whose purposes are indicated in the schema: accountability, delegation, monitoring, direction setting, alignment, execution and control.
  - Evaluate/monitor—The phase in which it is ensured that the information resource continues to work properly, i.e., to be valuable. Activities in this phase may refer to keeping information up to date as well as other kinds of information management activities, e.g., enhancing, cleansing, merging, removing duplicate information data in data warehouses.
  - Update/dispose—The phase in which the information resource is discarded when it is no longer of use. Activities in this phase may refer to information archiving or destroying.
- **Good practices**—The concept of information is understood differently in different disciplines such as economics, communication theory, information science, knowledge management and information systems; therefore, there is no universally agreed-on definition regarding what information is. The nature of information can, however, be clarified through defining and describing its properties.

The following scheme is proposed to structure information's different properties: It consists of six levels or layers to define and describe properties of information. These six levels present a continuum of attributes, ranging from the physical world of information, where attributes are linked to information technologies and media for information capturing, storing, processing, distribution and presentation, to the social world of information use, comprehension and action. The following descriptions can be given to the layers and information attributes:

– Physical world layer—The world where all phenomena that can be empirically observed take place
  · Information carrier/media—The attribute that identifies the physical carrier of the information, e.g., paper, electric signals, sound waves
– Empiric layer—The empirical observation of the signs used to encode information and their distinction from each other and from background noise
  · Information access channel—The attribute that identifies the access channel of the information, e.g., user interfaces
– Syntactic layer—The rules and principles for constructing sentences in natural or artificial languages. Syntax refers to the form of information.
  · Code/language—Attribute that identifies the representational language/format used for encoding the information and the rules for combining the symbols of the language to form syntactic structures.
– Semantic layer—The rules and principles for constructing meaning out of syntactic structures. Semantics refers to the meaning of information.
  · Information type—The attribute that identifies the kind of information, e.g., financial vs. non-financial information, internal vs. external origin of the information, forecasted/predicted vs. observed values, planned vs. realised values
  · Information currency—The attribute that identifies the time horizon referred to by the information, i.e., information on the past, the present or the future
  · Information level—The attribute that identifies the degree of detail of the information, e.g., sales per year, quarter, month
– Pragmatic layer—The rules and structures for constructing larger language structures that fulfil specific purposes in human communication. Pragmatics refers to the use of information.
  · Retention period—The attribute that identifies how long information can be retained before it is destroyed
  · Information status—The attribute that identifies whether the information is operational or historical
  · Novelty—The attribute that identifies whether the information creates new knowledge or confirms existing knowledge, i.e., information vs. confirmation
  · Contingency—The attribute that identifies the information that is required to precede this information (for it to be considered as information)
– Social world layer—The world that is socially constructed through the use of language structures at the pragmatic level of semiotics, e.g., contracts, law, culture.
  · Context—The attribute that identifies the context in which the information makes sense, is used, has value, etc., e.g., cultural context, subject domain context

6.2 Risk Function Perspective: Risk Governance- and Management-related Information

The purpose of this section is to identify and discuss all information items that are required to build and sustain effective and efficient risk governance and management in an enterprise. In other words, it lists information items that will support the risk, e.g., risk assessment, function.

Figure 28 lists a number of information items that form risk-related information sources for the enterprise.

Figure 28—Information Items Supporting Risk Governance and Management	
Information Item	**Definition/Description**

Risk profile	A risk profile is a description of the overall identified risk to which the enterprise is exposed. A risk profile consists of:
	• Risk register
	  – Risk scenario
	  – Risk analysis
	• Risk action plan
	• Loss events (historical and current)
	• Risk factor
	• Independent assessment findings

Risk register (or risk universe) [part of risk profile]	A risk register provides detailed information on: • Each identified risk, such as risk owner, details of the scenario and assumptions • Affected stakeholders • Causes/indicators • Information on the detailed scores, i.e., risk ratings, on the risk analysis • Detailed information on the risk response (e.g., action owner) and the risk response status (e.g., time frame for action) • Related projects • Risk tolerance level This can also be defined as the risk universe.
Risk scenario [part of risk register]	A risk scenario is a detailed description of an IT-related risk that can lead to a business impact, when it occurs. It includes elements such as: • Actor • Threat type • Event • Assets/resource • Time These elements are explained in more detail in section 2B, chapter 2.
Risk analysis results [part of risk register]	Risk scenario analysis is a technique to make risk more understandable and to allow for proper risk analysis and assessment. The risk analysis results consist of estimated scenario frequency and impact, loss forms, and options to reduce scenario frequency and impact.
Risk action plan [part of risk profile]	The risk action plan provides a framework that documents the priority order in which individual risk actions shall be implemented and how they shall be implemented. An action plan should be discussed with appropriate stakeholders and clearly document: • Risk scenarios that will be mitigated by the identified actions • Root cause of the scenario (root cause analysis [RCA]) • The reasons for selection of action options based on the control evaluation criteria • Those who are accountable for approving the plan and those who are responsible for implementing the plan • Proposed actions • Resource requirements, including contingencies • Performance measures and constraints • Cost vs. risk reduction benefit • Reporting and monitoring requirements • Timing and schedule
Loss event [part of risk profile]	A risk event that resulted in a loss. For example, error in systems processing/timing of trades can result in a loss or negative position. Include information on root cause and actual cost of the event, ideally broken up by the enterprise's standard loss forms.
Risk factor [part of risk profile]	A risk factor is a condition that can influence the frequency and impact and, ultimately, the business impact of IT-related events/scenarios. Risk factors can also be interpreted as causal factors of the scenario that is materialising, due to vulnerabilities or weaknesses. These factors include: • External context • Internal context • Risk management capabilities • IT-related capabilities
Independent assessment findings [part of risk profile]	Audit report that contains findings of assessments which should be taken into consideration during risk identification and analysis activities.

Figure 28—Information Items Supporting Risk Governance and Management *(cont.)*	
Information Item	**Definition/Description**
Risk communication plan	A risk communication plan is used to define frequency, types and recipients of information about risk. The main purpose of the plan is to reduce the overload of non-relevant information (avoiding 'risk noise').
Risk report	A risk report covers information on the current risk profile including risk map, risk dashboard, actual status of risk response, and emerging risk and trends. Further, it identifies the top ten risk items, which may be strategic, tactical or operational, that require management focus. This report is tailored to the requirements of the recipient.
Risk awareness programme	A risk awareness programme is a clearly and formally defined plan, structured approach and set of related activities and procedures with the objective of realizing and maintaining a risk-aware culture.
Risk map	A common, very easy and intuitive technique to present risk is the risk map. Risk is plotted on a two-dimensional diagram, with frequency and impact as the two dimensions. The risk map representation is powerful and provides an immediate and complete view on risk and apparent areas for action. Furthermore, a risk map allows defining colour zones that indicate appetite bands of significance in graphical mode.
Risk universe	The risk universe is all risk related to an enterprise, including the unknowns,[4] which could have an impact, either positively or negatively, on the ability of an enterprise to achieve its long term mission (or vision).
Risk appetite	Risk appetite is the broad-based amount of risk in different aspects that an enterprise is willing to accept in pursuit of its mission (or vision).
Risk tolerance[5]	Risk tolerance is the acceptable level of variation that management is willing to allow for any particular risk as it pursues objectives.
Key risk indicator (KRIs)	A risk indicator is a metric capable of showing that the enterprise is subject to, or has a high probability of being subject to, a risk that exceeds the defined risk appetite. A KRI is differentiated as being highly relevant and possessing a high probability of predicting or indicating important risk.
Emerging risk issues and factors	These consist of information on upcoming or likely combinations of controls, value and threat conditions that constitute a noteworthy level of future IT risk.
Risk taxonomy	Risk taxonomy is about providing a clear understanding of terminologies and scales to be used among the stakeholders while discussing and communicating risk. The taxonomy should be communicated and used enterprisewide.
Business impact analysis (BIA) report	This is a report resulting from the BIA, whose purpose it is to develop a common understanding of the business processes that are specific to each business unit, qualify the impact in the event of risk occurrence and critical to the survival of an enterprise.
Risk event	A risk event is something that happens at a specific place and/or time that can affect the proper business functions. Risk events can be broken down into threat events, loss events and vulnerability events.
Risk and control activity matrix	The risk and control activity matrix is a document that contains identified risk items, their ranking and control activities, and their design and operating effectiveness.
Risk assessment	A risk assessment is the process used to identify and qualify or quantify risk and its potential effects.

---

[4] Unknown risk is risk to which enterprises do not know they are susceptible. This definition is compatible with the COSO ERM definitions, which are equivalent to the ISO 31000 definition in ISO Guide 73.
[5] This definition is compatible with the COSO ERM definitions, which are equivalent to the ISO 31000 definition in ISO Guide 73.

Page intentionally left blank

CHAPTER 7
ENABLER: SERVICES, INFRASTRUCTURE AND APPLICATIONS

This chapter contains guidance on how services, infrastructure and applications items can enable risk governance and management provisioning in the enterprise (the risk function perspective). To that purpose, the following items are discussed:
• The Services, Infrastructure and Applications model
• A list of selected services relevant to the provisioning of risk governance and management
• A description of the detailed information that is provided for each service. The actual detailed information is included in appendix B.6.

7.1 The Services, Infrastructure and Applications Model

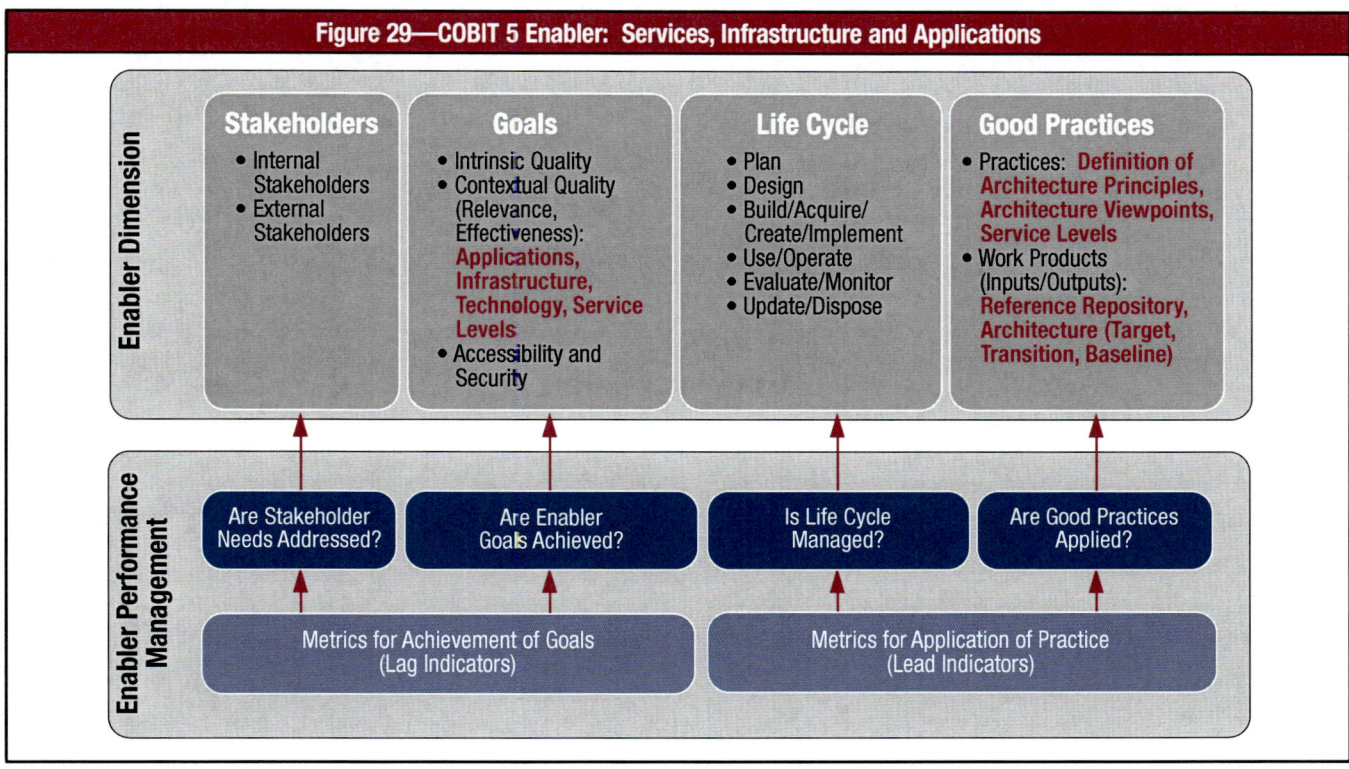

Figure 29—COBIT 5 Enabler: Services, Infrastructure and Applications

The Services, Infrastructure and Applications model (**figure 29**) shows:
• **Stakeholders**—Service capabilities (the combined term for services, infrastructure and applications) stakeholders can be internal and external. Services can be delivered by internal or external parties—internal IT departments, operations managers, outsourcing providers. Users of services can also be internal— business users—and external to the enterprise—partners, clients, suppliers. The stakes of each of the stakeholders need to be identified and are focused on either delivering adequate services or on receiving requested services from providers.
• **Goals**—Goals of the service-level capability are expressed in terms of services—applications, infrastructure, technology—and service levels, considering which services and service levels are most economical for the enterprise. Again, goals will relate to the services and how they are provided, as well as their outcomes, i.e., contribution towards successfully supported business processes.
• **Life cycle**—Service capabilities have a life cycle. The future or planned service capabilities are typically described in a target architecture. It covers the building blocks, such as future applications and the target infrastructure model, and also describes the linkages and relationships amongst these building blocks.

The current service capabilities that are used or operated to deliver current IT services are described in a baseline architecture. Depending on the time frame of the target architecture, a transition architecture may also be defined, which shows the enterprise at incremental states between the target and baseline architectures.

- **Good practices**—Good practice for service capabilities includes:
  - Definition of architecture principles—Architecture principles are overall guidelines that govern the implementation and use of IT-related resources within the enterprise. Examples of potential architecture principles are:
    - Reuse—Common components of the architecture should be used when designing and implementing solutions as part of the target or transition architectures.
    - Buy vs. build—Solutions should be purchased unless there is an approved rationale for developing them internally.
    - Simplicity—The enterprise architecture should be designed and maintained to be as simple as possible while still meeting enterprise requirements.
    - Agility—The enterprise architecture should incorporate agility to meet changing business needs in an effective and efficient manner.
    - Openness—The enterprise architecture should leverage open industry standards.
  - Definition of architecture viewpoints—The enterprise's definition of the most appropriate architecture viewpoints to meet the needs of different stakeholders. These are the models, catalogues and matrices used to describe the baseline, target or transition architectures; for example, an application architecture could be described through an application interface diagram, which shows the applications in use (or planned) and the interfaces amongst them.
  - Architecture repository—Having an architecture repository that can be used to store different types of architectural outputs, including architecture principles and standards, architecture reference models, and other architecture deliverables, and defines the building blocks of services such as:
    - Applications, providing business functionality
    - Technology infrastructure, including hardware, system software and networking infrastructure
    - Physical infrastructure
  - Service levels that are defined and achieved by the service providers

  External good practices for architecture frameworks and service capabilities exist. These are guidelines, templates or standards that can be used to fast track the creation of architecture deliverables. Examples include:
  - TOGAF (*www.opengroup.org/togaf*) provides a Technical Reference Model and an Integrated Information Infrastructure Reference Model.
  - ITIL provides comprehensive guidance on how to design and operate services.

7.2 Risk Function Perspective: Risk Governance- and Management-related Services, Infrastructure and Applications

The purpose of this section is to identify and discuss all services, infrastructure and applications that are required to build and sustain effective and efficient risk management in an enterprise. In other words, it lists all services and supporting applications that support the risk management function.

Services are a set of functions that the risk management process needs to provide, including stakeholder engagement, risk identification, risk analysis, risk reporting and risk prioritisation. **Figure 30** identifies and describes services, infrastructure and applications that the risk management process provides.

Figure 30—Risk Management-related Services	
Services and Supporting Applications	**Description**
Services	
Programme/project risk advisory services	The function that helps to ensure that new/changing business strategy, processes, and technology maintain an optimized level of risk
Incident management services	The function that helps the enterprise cost-effectively manage losses that may materialize from incidents
Architecture advisory services	The function that provides subject matter expert guidance to help ensure business, data and technology architecture supports the risk management objectives of the enterprise
Risk intelligence services	The function that provides both tactical and strategic threat, vulnerability and asset intelligence to support enterprise risk decisions
Risk management services	The function that provides risk subject matter expertise to support the development and ongoing support of risk management programmes within the enterprise
Crisis management services	The set of people, organisations, processes and technology that helps to respond to any type of crisis, including those requiring the activation of the business continuity plan (BCP)

Figure 30—Risk-Management-related Services *(cont.)*	
Services and Supporting Applications	**Description**
Infrastructure	
Data sources	The resources that provide timely risk environment data that support data mining and risk analytics
Infrastructure for knowledge repositories	The resources needed to store intelligence from internal and external source, e.g., big data
Intelligence integration architecture	The resources that provide a means of bridging real-time data sources, e.g., System Insight Manager (SIM) providers; knowledge repositories, e.g., ERM tools; and risk analytic applications to provide timely decision support
Applications	
Governance, risk and compliance (GRC) tools	A subset of GRC tools that enable the enterprise to collect, analyse, manage and report risk, including potential dashboards or balanced scorecard (BSC), as defined by the enterprise

These tools aim to communicate the risk in a prioritised order so that the core information can be extracted at a single glance. Risk matrix (risk map) is one such tool, which enables the enterprise to recognise the most critical risk items in the repository and how far out of risk appetite they are. |
Analysis tools	Qualitative and/or quantitative tools to support well-informed risk decision making
Tools for risk communication/reporting	These tools aim to communicate the findings of risk management
Knowledge repositories	A set of repositories to manage information used to facilitate the risk management analysis and overall process
Business continuity tools	These tools aim to help manage risk analysis, for example, BIA, and related activities in a risk management context.

Page intentionally left blank

CHAPTER 8
ENABLER: PEOPLE, SKILLS AND COMPETENCIES

This chapter contains guidance on how people, skills and competencies can enable risk governance and management in the enterprise (the risk function perspective). To that purpose, the following items are discussed:
• The People, Skills and Competencies model
• A list of selected skill sets relevant to the provisioning of risk governance and management
• A description of the skill sets that are deemed relevant for a risk analyst and a risk manager. This detailed information is included in appendix B.7.

8.1 The People, Skills and Competencies Model

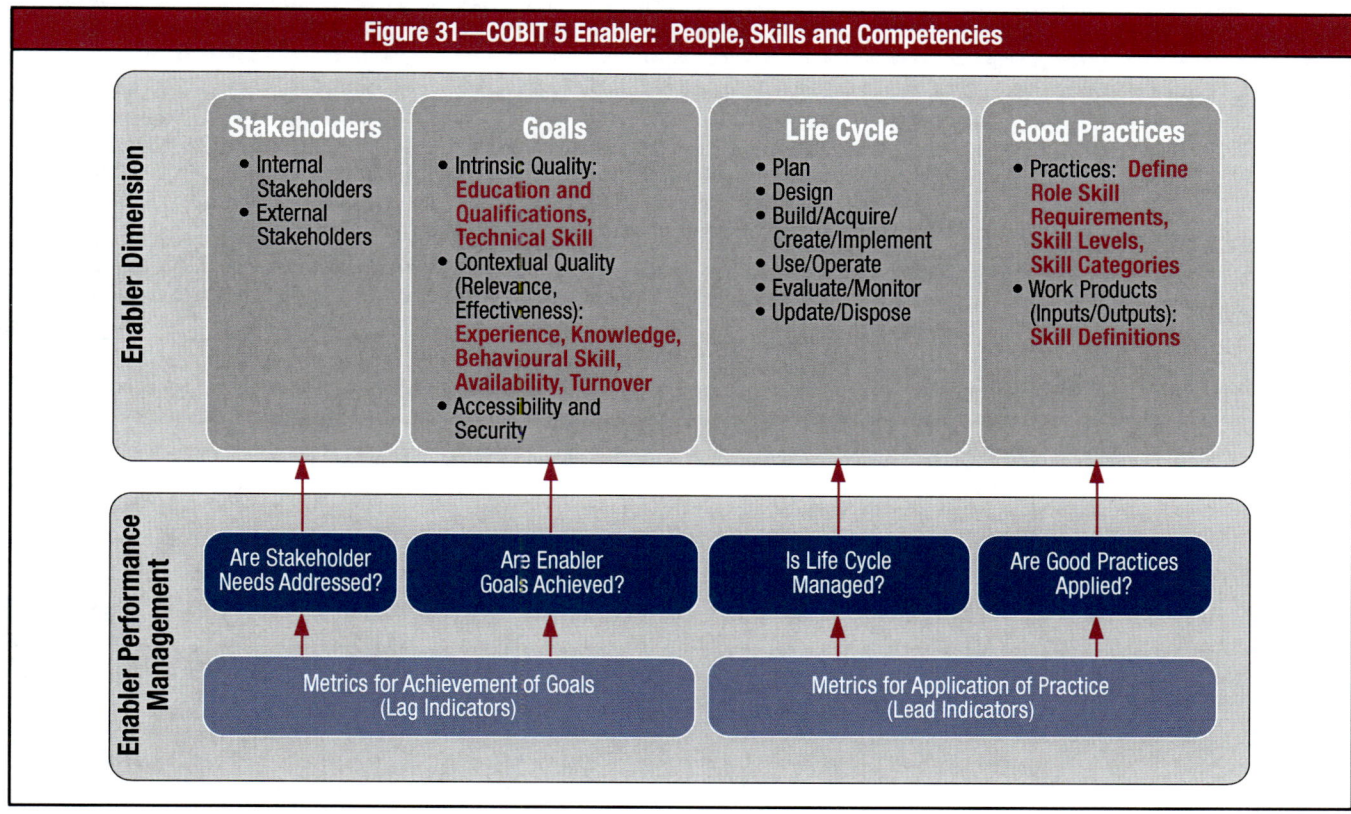

Figure 31—COBIT 5 Enabler: People, Skills and Competencies

The People, Skills and Competencies model (**figure 31**) shows:
• **Stakeholders**—Skills and competencies stakeholders are internal and external to the enterprise. Different stakeholders assume different roles—business managers, project managers, partners, competitors, recruiters, trainers, developers, technical IT specialists, etc.—and each role requires a distinct skill set. This section discusses the skill sets of a risk analyst and a risk manager.
• **Goals**—Goals for skills and competencies relate to education and qualification levels, technical skills, experience levels, knowledge and behavioural skills required to provide and perform successfully process activities, organisational roles, etc. Goals for people include correct levels of staff availability and turnover rate.
• **Life cycle**:
  – Skills and competencies have a life cycle. An enterprise has to know what its current skill base is and plan what it needs to be. This is influenced by (amongst other issues) the strategy and goals of the enterprise. Skills need to be developed (e.g., through training) or acquired (e.g., through recruitment) and deployed in the various roles within the organisational structure. Skills may need to be disposed of, e.g., if an activity is automated or outsourced.
  – Periodically, such as on an annual basis, the enterprise needs to assess the skill base to understand the evolution that has occurred, which will feed into the planning process for the next period.
  – This assessment can also feed into the reward and recognition process for human resources.
• **Good practices**:
  – Good practice for skills and competencies includes defining the need for objective skill requirements for each role played by the various stakeholders. This can be described through different skill levels in different skill categories. For each appropriate skill level in each skill category, a skill definition should be available. The skill categories correspond with the IT-related activities undertaken, e.g., information management, business analysis.

8.2 Risk Function Perspective: Risk Governance- and Management-related Skills and Competencies

The purpose of this section is to identify and describe all skill sets and competencies that are required to build and sustain effective and efficient risk management in an enterprise. In other words, it lists all skill sets and competencies needed to support the risk function (**figure 32**).

Figure 32—Risk Management Skill Sets and Competencies	
Skill Sets and Competencies	**Description**
Leadership skills	Risk management often involves many different stakeholders with differing opinions and values that must be navigated to drive effective business outcomes, requiring effective risk management leadership. Leadership skills include proactive leadership that sets clear direction that is aligned to the business outcomes and determination to ensure that the implemented policies deliver the effective disposition of risk. Leadership skills also require the ability to effectively work with all stakeholders to demonstrate effective escalation and communication. Lastly, risk needs to be managed in a cost-effective manner.
Analytical capability	The increasing complexity of business and the need to comply with regulations for each vertical industry requires that risk analysts have the capabilities to break down risk into risk factors that may prevent the achievement of goals and to assess those risk factors. This increasing complexity of business, legislative and regulatory requirements demands analytical skills to decipher them. To be able to analyse the entire risk as pieces to see where the fault lies and, then, to be able to put the pieces together in a useful and understandable manner, requires the risk analyst to approach the topic methodically and o have a structured mindset.
Critical thinking	Ability to make professional judgments about the value of additional information and determine whether a sufficient level of analysis has occurred is necessary. Ability to document and qualify assumptions and to articulate risk scenarios is also necessary.
Interpersonal capabilities	A key attribute of risk professionals is their ability to obtain information that is timely and accurate and to communicate with stakeholders who have different backgrounds and objectives. The risk professional uses non-technical language effectively, so that the message is meaningful to all stakeholders and demonstrates an essential understanding of business goals and priorities.
Communication	Risk needs to be communicated to stakeholders who have different backgrounds and objectives. The risk manager or analyst should have the capability to communicate risk, risk factors, and the associated loss exposure in the context, language and priority of the relevant stakeholder. The risk manager or analyst is capable of engaging all stakeholders in a meaningful way, using consistent nomenclature and providing examples for context.
Influencing	Risk professionals require well-developed persuasion skills to help with adoption of risk practices across the enterprise and demonstrate value to stakeholders.
Lateral thinking (thinking outside the box)	Risk needs to be approached differently depending on the type of risk. Ideas and techniques from other disciplines should be leveraged.
Technical understanding	The level of technical skills and competencies depends on the role within the risk function. To understand the vulnerabilities of IT systems and the threats that exploit them, one needs to have a basic understanding of the components comprising IT systems and how these components are connected to each other physically and logically.
Organisational and business awareness	To enable the enterprise to effectively plan, communicate and execute its risk management processes, the organisational points of contact, business units, goals, employee roles and responsibilities, and escalation paths must be documented and kept up to date.
Risk expertise	Expertise in threat sources, threat scenarios, vulnerabilities and impact on the business is critical to success. This skill refers to an understanding of the basic nature and composition of risk as well as ongoing improvement to keep pace with the dynamic nature of threats, vulnerabilities and impacts in the modern business environment.
Training and coaching	Risk management is a part of all roles in the enterprise. To confirm appropriate levels of risk expertise and practice, enterprises require capabilities in effective training and coaching of stakeholders. The training programmes must allow for training based on varying levels of risk awareness. The ability to deliver targeted training programmes is essential in the successful update and sustainability of risk practices. Training programmes must be provided to ensure that individuals are effective within their roles.

These are all general skills relevant to risk management within the enterprise. Each function or role requires a combination of several of these skills. As an example, the detailed profiles of a risk analyst and a risk manager are presented in appendix B.7. These examples should be interpreted as general guidance on common experience, knowledge and expertise for these profiles. They are not specific requirements to the role.

SECTION 2B. THE RISK MANAGEMENT PERSPECTIVE AND USING COBIT 5 ENABLERS

This section comprises:
- The core risk management processes used to implement effective and efficient risk management for the enterprise to support stakeholder value
- Risk scenarios, i.e., the key information item needed to identify, analyse and respond to risk. Risk scenarios are the tangible and assessable representation of risk.
- How COBIT 5 enablers can be used to respond to unacceptable risk scenarios

CHAPTER 1
CORE RISK PROCESSES

The core processes for risk governance and risk management are described in the COBIT 5 processes EDM03 *Ensure risk optimisation* and APO12 *Manage risk* (**figure 33**). These processes comprise the core activities of the risk function that is discussed in section 2A. They support the enterprise in obtaining stakeholder value and enterprise objectives while optimising resources and risk.

Figure 33—Core Risk Processes	
COBIT 5 Process Identification	**Reasoning**
EDM03 Ensure Risk Optimisation	This process covers the understanding, articulation and communication of the enterprise risk appetite and tolerance and ensures identification and management of risk to the enterprise value that is related to IT use and its impact. The goals of this process are to: • Define and communicate risk thresholds and make sure that key IT-related risk is known. • Effectively and efficiently manage critical IT-related enterprise risk. • Ensure IT-related enterprise risk does not exceed risk appetite.
APO12 Manage Risk	This process covers the continuous identification, assessment and reduction of IT-related risk within levels of tolerance set by enterprise executive management. Management of IT-related enterprise risk should be integrated with overall ERM. The costs and benefits of managing IT-related enterprise risk should be balanced by: • Collecting appropriate data and analysing risk • Maintaining the risk profile of the enterprise and articulating risk • Defining the risk management action portfolio and responding to risk

These core processes and their activities are described more in detail in appendix C.

Page intentionally left blank

Page intentionally left blank

CHAPTER 2
RISK SCENARIOS

2.1 Introduction

A key information item used in the core risk management process APO12 is the risk scenario (**figure 34**).

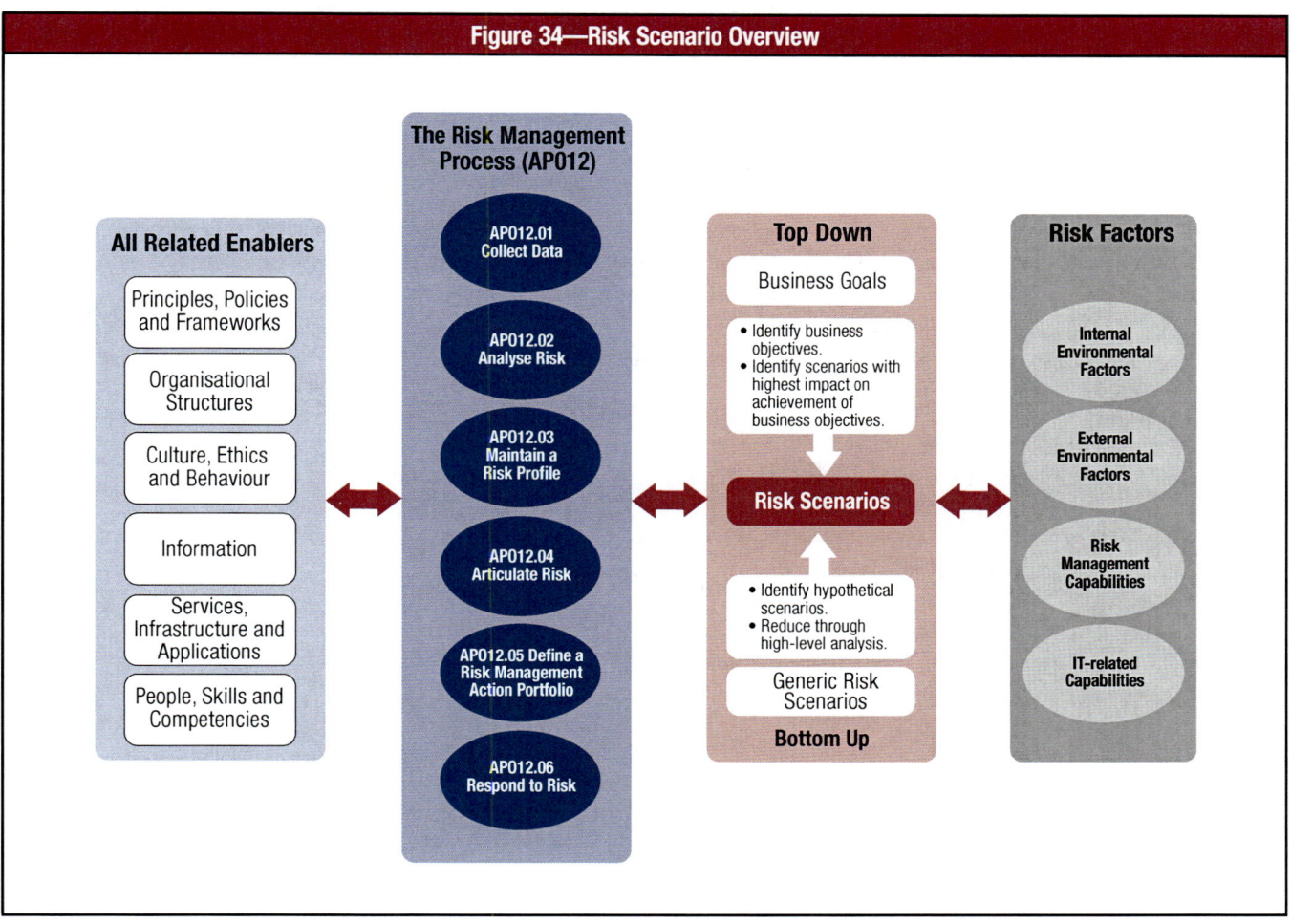

Figure 34—Risk Scenario Overview

A **risk scenario** is a description of a possible event that, when occurring, will have an uncertain impact on the achievement of the enterprise's objectives. The impact can be positive or negative.

The core risk management process requires risk needs to be identified, analysed and acted on. Well-developed risk scenarios support these activities and make them realistic and relevant to the enterprise.

Figure 34 also shows that risk scenarios can be derived via two different mechanisms:
• A top-down approach, where one starts from the overall enterprise objectives and performs an analysis of the most relevant and probable IT risk scenarios impacting the enterprise objectives. If the impact criteria used during risk analysis are well aligned with the real value drivers of the enterprise, relevant risk scenarios will be developed.
• A bottom-up approach, where a list of generic scenarios is used to define a set of more relevant and customised scenarios, applied to the individual enterprise situation.

The approaches are complementary and should be used simultaneously. Indeed, risk scenarios must be relevant and linked to real business risk. On the other hand, using a set of example generic risk scenarios could assist to identify risk and reduce the chance of overlooking major/common risk scenarios and can provide a comprehensive reference for IT risk. However, specific risk items for each enterprise and critical business requirements need to be considered in the enterprise risk scenarios.

Note: Do not over rely on the list of example generic risk scenarios. The list, although quite comprehensive, broad and covering most potential risk items, needs to be adapted to the enterprise specific situation. It is not intended that, going forward, all IT risk management will use the same set of predefined IT risk scenarios. Rather, it is encouraged that this list be used as a basis for the development of specific, relevant scenarios.

2.2 Developing Risk Scenarios Workflow

In practice, the following approach is suggested:
• Use the list of example generic risk scenarios (see **figure 38** in chapter 3) to define a manageable set of tailored risk scenarios for the enterprise. To determine a manageable set of scenarios a business might begin by considering commonly occurring scenarios in its industry or product area, scenarios representing threat sources that are increasing in number or severity, and scenarios that involve legal and regulatory requirements applicable to the business. Another approach might be to identify high-risk business units and assess one or two high-risk operating processes within each, including the IT components that enable that process. Also, some less common situations should be included in the scenarios.
• Perform a validation against the business objectives of the entity. Do the selected risk scenarios address potential impacts on achievement of business objectives of the entity, in support of the overall enterprise's business objectives?
• Refine the selected scenarios based on this validation; detail them to a level in line with the criticality of the entity.
• Reduce the number of scenarios to a **manageable set**. 'Manageable' does not signify a fixed number, but should be in line with the overall importance (size) and criticality of the unit. There is no general rule, but if scenarios are reasonably and realistically scoped, the enterprise should expect to develop at least a few dozen scenarios.
• Keep all scenarios in a list so they can be re-evaluated in the next iteration and included for detailed analysis if they have become relevant at that time.
• Include in the scenarios an unspecified event, e.g., an incident not covered by other scenarios.

Once the set of risk scenarios is defined, it can be used for risk analysis, where frequency and impact of the scenario are assessed. Important components of this assessment are the risk factors.

The enterprise can also consider evaluating scenarios that have a chance of occurring simultaneously. This is frequently referred to as 'stress' testing and actually entails combining multiple scenarios and understanding what the extra impact would be of them occurring together.

2.3 Risk Factors

Risk factors are those conditions that influence the frequency and/or business impact of risk scenarios. They can be of different natures and can be classified into two major categories:
• **Contextual factors**—Can be divided into internal and external factors, the difference being the degree of control an enterprise has over them:
  – Internal contextual factors—To a large extent, are under the control of the enterprise, although they may not always be easy to change
  – External contextual factors–To a large extent, are outside the control of the enterprise
• **Capabilities**—How effective and efficient the enterprise is in a number of IT-related activities. They can be distinguished in line with the COBIT 5 framework:
  – IT risk management capabilities—Indicate to what extent the enterprise is mature in performing the risk management processes
  – IT-related capabilities—Indicate the capability of the IT-related COBIT 5 enablers

The importance of risk factors lies in the influence they have on the risk. They are heavy influencers on the frequency and impact of IT scenarios and should be taken into account during every risk analysis.

Risk factors can also be interpreted as causal factors of the scenario that is materialising, or as vulnerabilities or weaknesses. These are terms often used in other risk management frameworks.

Scenario analysis should not only be based on past experience and known current events, but should also look forward to possible future circumstances.

Figure 35 depicts risk factors, which are discussed in more detail in the following paragraphs.

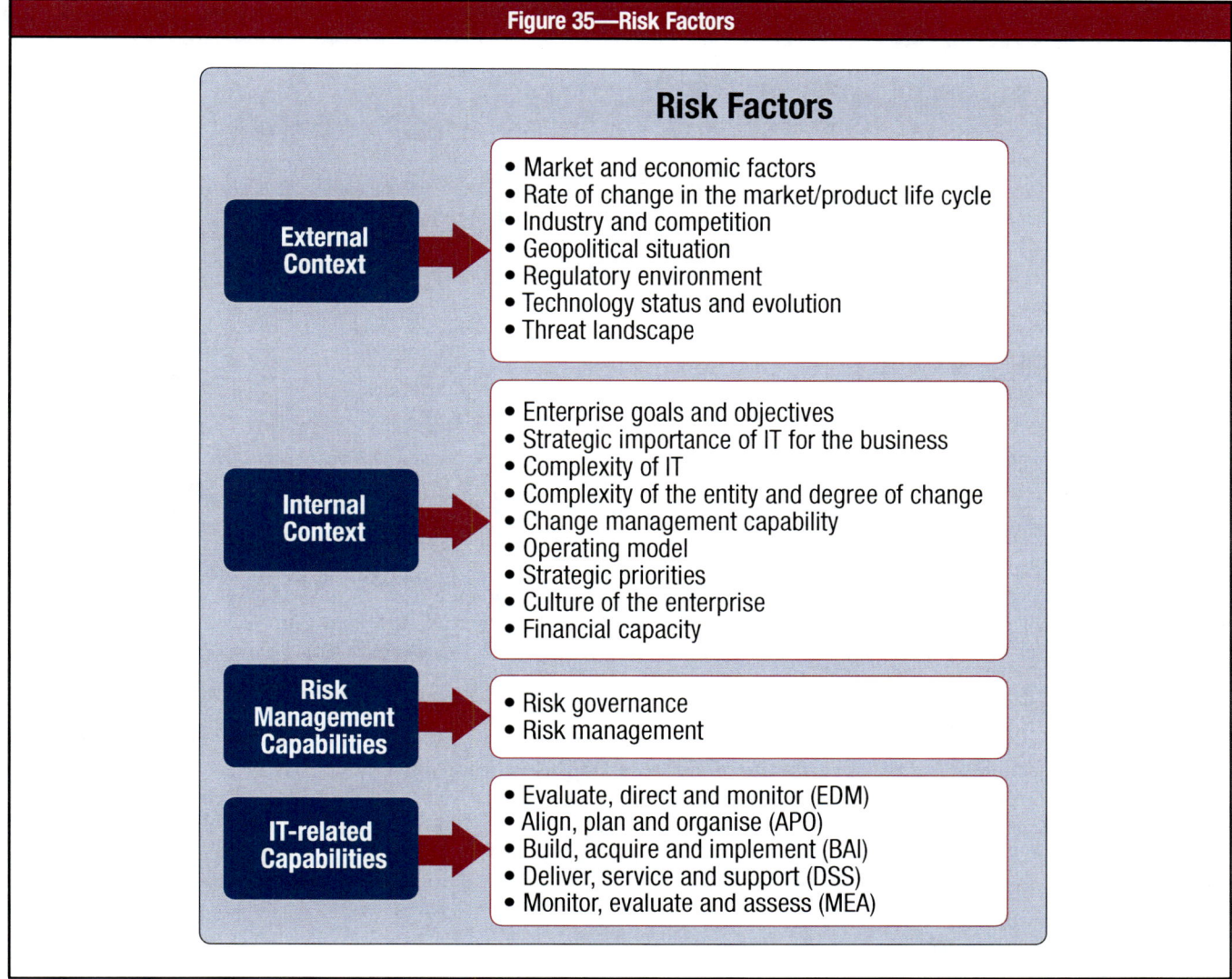

Figure 35—Risk Factors

Risk Factors

External Context
- Market and economic factors
- Rate of change in the market/product life cycle
- Industry and competition
- Geopolitical situation
- Regulatory environment
- Technology status and evolution
- Threat landscape

Internal Context
- Enterprise goals and objectives
- Strategic importance of IT for the business
- Complexity of IT
- Complexity of the entity and degree of change
- Change management capability
- Operating model
- Strategic priorities
- Culture of the enterprise
- Financial capacity

Risk Management Capabilities
- Risk governance
- Risk management

IT-related Capabilities
- Evaluate, direct and monitor (EDM)
- Align, plan and organise (APO)
- Build, acquire and implement (BAI)
- Deliver, service and support (DSS)
- Monitor, evaluate and assess (MEA)

External Context

Contextual IT risk factors, i.e., those circumstances that can increase the frequency or impact of an event and which are not always directly controllable by the enterprise, include:
- **Market/economic factors**—The industry sector in which the enterprise operates, i.e., operating in the financial sector requires different IT requirements and IT capabilities than operating in a manufacturing environment. Other economic factors can be included as well, e.g., nationalisation, mergers and acquisitions, consolidations.
- **Rate of change in the market in which the enterprise operates**—Are business models changing fundamentally? Is the product or service at the end of an important life cycle moment?
- **Competitive environment**—In which the enterprise operates
- **Geopolitical situation**—Is the geographic location subject to frequent natural disasters? Does the local political and overall economic context represent an additional risk?
- **Regulatory environment**—Is the enterprise subject to new or more strict IT-related regulations or regulations impacting IT? Are there any other compliance requirements beyond regulation, e.g., industry-specific, contractual?
- **Technology status and evolution**—Is the enterprise using state-of-the art technology and, more important, how fast are relevant technologies evolving?
- **Threat landscape**—How are relevant threats evolving in terms of frequency of occurring and level of capability?

Internal Context

Internal risk factors include:
- **Enterprise goals and objectives**—What are the needs of the stakeholders and how could these be impacted by risk?
- **Strategic importance of IT in the enterprise**—Is IT a strategic differentiator, a functional enabler or a supporting function?
- **Complexity of IT**—Is IT highly complex (e.g., complex architecture, recent mergers) or is IT simple, standardised and streamlined?

- **Complexity of the enterprise** (including geographic spread and value chain coverage, e.g., in a manufacturing environment)—Does the enterprise manufacture and distribute parts, and/or is it also doing assembly activities?
- **Degree of change**—What degree of changes is the enterprise is experiencing?
- **Change management capability**—To what extent is the enterprise capable of organisational change?
- **The risk management philosophy**—What is the risk philosophy of the enterprise (risk averse or risk taking) and, linked with that, the values of the enterprise?
- **Operating model**—The degree to which the enterprise operates independently or is connected to its clients/suppliers, the degree of centralisation/decentralisation
- **Strategic priorities**—What are the strategic priorities of the enterprise?
- **Culture of the enterprise**—Does the existing culture of the enterprise require changing to be able to effectively embrace risk management?
- **Financial capacity**—The capacity of the enterprise to provide financial support to enhance and maintain the IT environment whilst optimising risk

Risk Management Capability
Risk management capability is an indication of how well the enterprise is executing the core risk management processes and the related enablers. This can be measured by using a risk scorecard. The better performing the enablers are, the more capable the risk management programme is.

This factor is correlated with the capability of the enterprise to recognise and detect risk and adverse events; hence, it should not be neglected.

Risk management capability is a very significant element in the frequency and impact of risk events in an enterprise because it is responsible for management's risk decisions (or lack thereof), as well as for the presence, absence and/or effectiveness of controls that exist within an enterprise.

IT-related Capability
IT-related capabilities are associated with the capability level of IT processes and all other enablers. The generic enabler model in COBIT 5 contains an enabler performance model supporting capability assessments. A high maturity with regard to the different enablers is equivalent to high IT related capabilities, which can have a positive influence on:
- Reducing the frequency of events, e.g., having good software development processes in place to deliver high-quality and stable software, or having good security measures in place to reduce the number of security-related incidents
- Reducing the business impact when events happen, e.g., having a good BCP/DRP in place when disaster strikes

2.4 IT Risk Scenario Structure

An IT risk scenario is a description of an IT-related event that can lead to a business impact, when and if it should occur. For risk scenarios to be complete and usable for risk analysis purposes, they should contain the following components, as shown in **figure 36**:
- **Actor**—Who generates the threat that exploits a vulnerability. Actors can be internal or external and they can he human or non-human:
  - Internal actors are within the enterprise, e.g., staff, contractors.
  - External actors include outsiders, competitors, regulators and the market.

  Not every type of threat requires an actor, e.g., failures or natural causes.
- **Threat type** (the nature of the event)—Is it malicious? If not, is it accidental or is it a failure of a well-defined process? Is it a natural event?
- **Event**—Is it disclosure of confidential information, interruption of a system or of a project, theft or destruction? Action also includes ineffective design of systems, processes, etc., inappropriate use, changes in rules and regulation that will materially impact a system) or ineffective execution of processes, e.g., change management procedures, acquisition procedures, project prioritisation processes.
- **Asset/resource**—On which the scenario acts. An asset is any item of value to the enterprise that can be affected by the event and lead to business impact. A resource is anything that helps to achieve IT goals. Assets and resources can be identical, e.g., IT hardware is an important resource because all IT applications use it, and at the same time, it is an asset because it has a certain value to the enterprise. Assets/resources include:
  - People and skills
  - Organisational structures
  - IT processes, e.g., modelled as COBIT 5 processes, or business processes
  - Physical infrastructure, facilities, equipment, etc.
  - IT infrastructure, including computing hardware, network infrastructure, middleware
  - Other enterprise architecture components, including information, applications

Assets can be critical or not, e.g., a client-facing web site of a major bank compared to the web site of the local garage or the intranet of the software development group. Critical resources will probably attract a greater number of attacks or greater attention on failure; hence, the frequency of related scenarios will probably be higher. It takes skill, experience and thorough understanding of dependencies to understand the difference between a critical asset and a non-critical asset.

• **Time**—Dimension, where the following could be described, if relevant to the scenario:
  – The duration of the event, e.g., extended outage of a service or data centre
  – The timing (Does the event occur at a critical moment?)
  – Detection (Is detection immediate or not?)
  – Time lag between the event and consequence (Is there an immediate consequence, e.g., network failure, immediate downtime, or a delayed consequence, e.g., wrong IT architecture with accumulated high costs over a time span of several years?)

It is important to stay aware of the differences between loss events, threat events and vulnerability events. When a risk scenario materialises, a loss event occurs. The loss event has been triggered by a threat event (threat type plus event in **figure 36**. The frequency of the threat event leading to a loss event is influenced by the risk factors or vulnerability. Vulnerability is usually a state and can be increased/decreased by vulnerability events, e.g., the weakening of controls or by the threat strength. **One should not mix these three types of events into one big 'risk list'.**

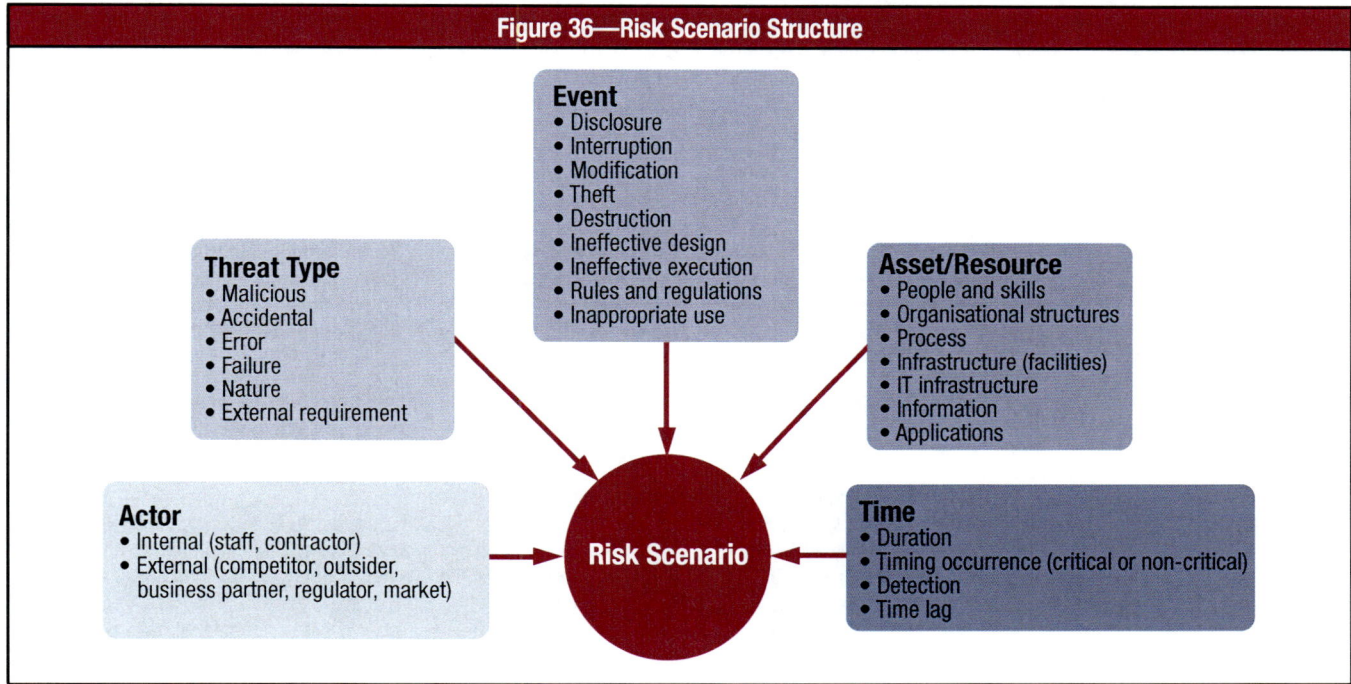

Figure 36—Risk Scenario Structure

Chapter 3 contains a set of generic IT risk scenarios that are built in line with the model described in the previous paragraphs. The set of generic scenarios contains examples of negative outcomes, but also examples where a risk, when managed well, can lead to a positive outcome.

2.5 Main Issues When Developing and Using Risk Scenarios

The use of scenarios is key to risk management, and the technique is applicable to any enterprise. Each enterprise needs to build a set of scenarios (containing the components described previously) as a starting point to conduct its risk analysis.

Building a complete set of scenarios means—in theory—that each possible value of every component should be combined. Each combination should then be assessed for relevance and realism and, if found to be relevant, entered into the risk register. In practice, this is not possible; very quickly, an unfeasible number of different risk scenarios can be generated. The number of scenarios to be developed and analysed should be kept to a relatively small number in order to remain manageable.

Figure 37 shows some of the main areas of focus/issues to address when using the risk scenario technique.

Figure 37—Risk Scenario Technique Main Focus Areas	
Focus/Issue	**Summary Guidance**
Maintain currency of risk scenarios and risk factors.	Risk factors and the enterprise change over time; hence, scenarios will change over time, over the course of a project or over the evolution of technology. For example, it is essential that the risk function develop a review schedule and the CIO works with the business lines to review and update scenarios for relevance and importance. Frequency of this exercise depends on the overall risk profile of the enterprise and should be done at least on an annual basis, or when important changes occur.
Use generic risk scenarios as a starting point and build more detail where and when required.	One technique of keeping the number of scenarios manageable is to propagate a standard set of generic scenarios through the enterprise and develop more detailed and relevant scenarios when required and warranted by the risk profile only at lower (entity) levels. The assumptions made when grouping or generalising should be well understood by all and adequately documented because they may hide certain scenarios or be confusing when looking at risk response. For example, if 'insider threat' is not well defined within a scenario, it may not be clear whether this threat includes privileged and non-privileged insiders. The differences between these aspects of a scenario can be critical when one is trying to understand the frequency and impact of events, as well as mitigation opportunities.
Number of scenarios should be representative and reflect business reality and complexity.	Risk management helps to deal with the enormous complexity of today's IT environments by prioritising potential action according to its value in reducing risk. Risk management is about reducing complexity, not generating it; hence, another plea for working with a manageable number of risk scenarios. However, the retained number of scenarios still needs to accurately reflect business reality and complexity.
Risk taxonomy should reflect business reality and complexity.	There should be a sufficient number of risk scenario scales reflecting the complexity of the enterprise and the extent of exposures to which the enterprise is subject. Potential scales might be a 'low, medium, high' ranking or a numeric scale that scores risk importance from 0 to 5. Scales should be aligned throughout the enterprise to ensure consistent scoring.
Use generic risk scenario structure to simplify risk reporting	Similarly, for risk reporting purposes, entities should not report on all specific and detailed scenarios, but could do so by using the generic risk structure. For example, an entity may have taken generic scenario 15 (project quality), translated it into five scenarios for its major projects, subsequently conducted a risk analysis for each of the scenarios, then aggregated or summarised the results and reported back using the generic scenario header 'project quality'.
Ensure adequate people and skills requirements for developing relevant risk scenarios.	Developing a manageable and relevant set of risk scenarios requires: • Expertise and experience, to not overlook relevant scenarios and not be drawn into highly unrealistic[6] or irrelevant scenarios. While the avoidance of scenarios that are unrealistic or irrelevant is important in properly utilising limited resources, some attention should be paid to situations that are highly infrequent and unpredictable, but which could have a cataclysmic impact on the enterprise. • A thorough understanding of the environment. This includes the IT environment (e.g., infrastructure, applications, dependencies between applications, infrastructure components), the overall business environment, and an understanding of how and which IT environments support the business environment to understand the business impact. • The intervention and common views of all parties involved—senior management, which has the decision power; business management, which has the best view on business impact; IT, which has the understanding of what can go wrong with IT; and risk management, which can moderate and structure the debate amongst the other parties. • The process of developing scenarios usually benefits from a brainstorming/workshop approach, where a high-level assessment is usually required to reduce the number of scenarios to a manageable, but relevant and representative, number.
Use the risk scenario building process to obtain buy-in.	Scenario analysis is not just an analytical exercise involving 'risk analysts'. A significant additional benefit of scenario analysis is achieving organisational buy-in from enterprise entities and business lines, risk management, IT, finance, compliance and other parties. Gaining this buy-in is the reason why scenario analysis should be a carefully facilitated process.
Involve first line of defence in the scenario building process.	In addition to co-ordinating with management, it is recommended that selected members of the staff who are familiar with the detailed operations be included in discussions, where appropriate. Staff whose daily work is in the detailed operations are often more familiar with vulnerabilities in technology and processes that can be exploited.
Do not focus only on rare and extreme scenarios.	When developing scenarios, one should not focus only on worst-case events because they rarely materialise, whereas less-severe incidents happen more often.

---

[6] Unrealistic signifies not fixed in time or static. What used to be unthinkable, mainly because it never happened or because it happened too long ago, becomes realistic as soon as it occurs again. A striking example is the 11 September 2001 terrorist attacks in the US. It is human nature for things that have not yet happened, even when they are theoretically possible, to be estimated as not possible or extremely unlikely. Only when they occur will they be taken seriously in risk assessments. This may be regarded as lack of foresight or lack of due care, but it is actually the essence of risk management—trying to shape and contain the future based on past experience and future predictions.

Figure 37—Risk Scenario Technique Main Focus Areas *(cont.)*	
Focus/Issue	**Summary Guidance**
Deduce complex scenarios from simple scenarios by showing impact and dependencies.	Simple scenarios, once developed, should be further fine-tuned into more complex scenarios, showing cascading and/or coincidental impacts and reflecting dependencies. For example: • A scenario of having a major hardware failure can be combined with the scenario of failed DRP. • A scenario of major software failure can trigger database corruption and, in combination with poor data management backups, can lead to serious consequences, or at least consequences of a different magnitude than a software failure alone. • A scenario of a major external event can lead to a scenario of internal apathy.
Consider systemic and contagious risk.	Attention should be paid to systemic and/or contagious risk scenarios: • **Systemic**—Something happens with an important business partner, affecting a large group of enterprises within an area or industry. An example would be a nationwide air traffic control system that goes down for an extended period of time, e.g , six hours, affecting air traffic on a very large scale. • **Contagious**—Events that happen at several of the enterprise's business partners within a very short time frame. An example would be a clearinghouse that can be fully prepared for any sort of emergency by having very sophisticated disaster recovery measures in place, but when a catastrophe happens, finds that no transactions are sent by its providers and hence is temporarily out of business.
Use scenario building to increase awareness for risk detection.	Scenario development also helps to address the issue of detectability, moving away from a situation where an enterprise 'does not know what it does not know'. The collaborative approach for scenario development assists in identifying risk to which the enterprise, until then, would not have realised it was subject to (and hence would never have thought of putting in place any countermeasures). After the full set of risk items is identified during scenario generation, risk analysis assesses frequency and impact of the scenarios. Questions to be asked include: • Will the enterprise ever detect that the risk scenario has materialised? • Will the enterprise notice something has gone wrong so it can react appropriately? Generating scenarios and creatively thinking of what can go wrong will automatically raise and, hopefully, cause response to, the question of detectability. Detectability of scenarios includes two steps: visibility and recognition. The enterprise must be in a position that it can observe anything going wrong, and it needs the capability to recognise an observed event as something wrong.

Page intentionally left blank

CHAPTER 3
GENERIC RISK SCENARIOS

An IT risk scenario is a description of an IT-related event that can lead to a loss event that has a business impact, when and if it should occur. The generic scenarios serve, after customisation, as input to risk analysis activities, where the ultimate business impact (amongst others) needs to be established. This chapter contains a set of generic IT risk scenarios (**figure 38**), built in line with the model described in the previous section of this guide. The set of generic scenarios contains both negative and positive examples scenarios.

A word of warning: The table with generic scenarios does not replace the creative and reflective phase that every scenario-creating exercise should contain. In other words, it is not recommended that an enterprise blindly use this list and assume that no other risk scenarios are possible, or assume that every scenario contained in the list is applicable to the enterprise. Intelligence and experience are needed to derive a relevant and customised list of scenarios starting from this generic list.

The generic risk scenarios in **figure 38** include the following information:
• **Risk scenario category**—High-level description of the category of scenario (e.g., IT project selection). In total there are 20 categories
• **Risk scenario components**—Provides details about the threat type, actor, event, asset/resource and time of every scenario category
• **Risk type**—The type to which scenarios derived from this generic scenario will fit, using the three risk types explained earlier:
  – **IT benefit/value enablement risk**—Associated with (missed) opportunities to use technology to improve the efficiency or effectiveness of business processes or as an enabler for new business initiatives
  – **IT programme and project delivery risk**—Associated with the contribution of IT to new or improved business solutions, usually in the form of projects and programmes
  – **IT operations and service delivery risk**—Associated with the operational stability, availability, protection and recoverability of IT services, which can bring destruction or reduction of value to the enterprise

A 'P' indicates a primary (higher degree) fit and an 'S' represents a secondary (lower degree) fit. Blank cells indicate that the risk category is not relevant for the risk scenario at hand.
• **Example scenarios**—For each scenario category, one or several small examples are given of scenarios with a negative outcome, indicating whether it is more of a value destruction or a failure to gain, and/or positive outcome, indicating value gain. In total, **111 risk scenario examples** are included with possible negative and/or positive outcomes.

Figure 38—Example Risk Scenarios						
		Risk Type			Example Scenarios	
Ref.	Risk Scenario Category	IT Benefit/Value Enablement	IT Programme and Project Delivery	IT Operations and Service Delivery	Negative Example Scenarios	Positive Example Scenarios
0101	Portfolio establishment and maintenance	P	P	S	Wrong programmes are selected for implementation and are misaligned with corporate strategy and priorities.	Programmes lead to successful new business initiatives selected for execution.
0102		P	P	S	There is duplication between initiatives.	Aligned initiatives have streamlined interfaces.
0103		P	P	S	A new important programme creates long-term incompatibility with the enterprise architecture.	New programmes are assessed for compatibility with existing architecture.
0104		P	P	S	Competing resources are allocated and managed inefficiently and are misaligned to business priorities.	

		Risk Type			Example Scenarios	
Ref.	**Risk Scenario Category**	**IT Benefit/Value Enablement**	**IT Programme and Project Delivery**	**IT Operations and Service Delivery**	**Negative Example Scenarios**	**Positive Example Scenarios**
0201	Programme/projects life cycle management (programme/ projects initiation, economics, delivery, quality and termination)	P	P	S	Failing (due to cost, delays, scope creep, changed business priorities) projects are not terminated.	Failing or irrelevant projects are stopped on a timely basis.
0202		S	P	S	There is an IT project budget overrun.	The IT project is completed within agreed-on budgets.
0203		S	P		There is occasional late IT project delivery by an internal development department.	Project delivery is on time.
0204		P	P	S	Routinely, there are important delays in IT project delivery.	The project critical path is managed accordingly and delivery is on time.
0205		P	P	S	There are excessive delays in outsourced IT development project.	Communication with third parties ensures the timely delivery within agreed-on scope and quality.
0206		P	P		Programmes/projects fail due to not obtaining the active involvement throughout the programme/project life cycle of all stakeholders (including sponsor).	Change management is conducted appropriately throughout the life cycle of the programme/project to inform stakeholders on progress and train future users.
0301	IT investment decision making	P		S	Business managers or representatives are not involved in important IT investment decision making (e.g., new applications, prioritisation, new technology opportunities).	There is co-ordinated decision making over IT investments between business and IT.
0302		P		S	The wrong software, in terms of cost, performance, features, compatibility, etc., is selected for implementation.	Upfront analysis is performed and a business case is made up to ensure the adequate selection of software.
0303		P		P	The wrong infrastructure, in terms of cost, performance, features, compatibility, etc., is selected for implementation.	Upfront analysis is performed and a business case is made up to ensure the adequate selection of infrastructure.
0304		P	P		Redundant software is purchased.	
0401	IT expertise and skills	P	P	P	There is a lack of or mismatched IT-related skills within IT, e.g., due to new technologies.	Attracting the appropriate staff increases the service delivery of the IT department.
0402		P	P	P	There is a lack of business understanding by IT staff affecting the service delivery/ projects quality.	Correct staff and skill mix supports project delivery and value delivery.
0403		P	P	P	There are insufficient skills to cover the business requirements.	Correct skill mix and training ensures that there is a thorough understanding of the business by staff and allows full coverage of business requirements.
0404		S	P	P	There is an inability to recruit IT staff.	The correct amount of IT staff, with appropriate skills and competencies is attracted to support the business objectives.
0405		S	P	P	There is a lack of due diligence in the recruitment process.	Candidates are screened to ensure that appropriate skills, competencies and attitude are present.
0406		S	P	P	There is a lack of training leading to IT staff leaving.	IT staff members are able to determine their own training plan based on their aspirations and domains of interest, in collaboration with their superiors.
0407		S	P	P	There is insufficient return on investment regarding training due to early leaving of trained IT staff (e.g., MBA).	Career development is made formal and individual paths are determined to ensure IT staff is motivated to stay for a considerable amount of time.

Figure 38—Example Risk Scenarios *(cont.)*

		Risk Type			Example Scenarios	
Ref.	Risk Scenario Category	IT Benefit/Value Enablement	IT Programme and Project Delivery	IT Operations and Service Delivery	Negative Example Scenarios	Positive Example Scenarios
0408	IT expertise and skills *(cont.)*	S	P	P	There is an overreliance on key IT staff.	Job rotation ensures that nobody alone possesses the entire knowledge of the execution of a certain activity.
0409		S	P	P	There is an inability to update the IT skills to the proper level through training.	Training, attending seminars and reading thought leadership ensures that IT staff is up to date with the latest developments in its area of speciality.
0501	Staff operations (human error and malicious intent)	S	S	P	Access rights from prior roles are abused.	HR and IT administration co-ordinate on a frequent basis to ensure timely removal of access rights, avoiding the possibility of abuse.
0502		S		P	IT equipment is accidentally damaged by staff.	
0503		S		P	There are errors by IT staff (during backup, during upgrades of systems, during maintenance of systems, etc.).	The 4-eye principle is applied, decreasing the possibility of errors before moving to production.
0504		S		P	Information is input incorrectly by IT staff or system users.	The 4-eye principle is applied, decreasing the possibility of incorrect information input.
0505		S		P	The data centre is destroyed (sabotage, etc.) by staff.	Data centre is appropriately secured, only allowing access to authorised IT staff.
0506		S		P	There is a theft of a device with sensitive data by staff.	Office premises are secured and monitored for irregular activity.
0507		S		P	There is a theft of a key infrastructure component by staff.	Key infrastructure components are monitored 24/7 for performance, availability, etc. Alarm bells are raised in case of irregularities and acted on immediately.
0508		P	S	P	Hardware components were configured erroneously.	An enterprisewide configuration management system is set up, ensuring aligned configuration across the enterprise.
0509		P	S	P	Critical servers in the computer room were damaged (e.g., accident, etc.).	Key infrastructure components are monitored 24/7 for performance, availability, etc. Alarm bells are raised in case of irregularities and acted on immediately.
0510		P	S	P	Hardware was tampered with intentionally (security devices, etc.).	Key infrastructure components are monitored 24/7 for performance, availability, etc. Alarm bells are raised in case of irregularities and acted on immediately.

Figure 38—Example Risk Scenarios *(cont.)*

Ref.	Risk Scenario Category	Risk Type			Example Scenarios	
		IT Benefit/Value Enablement	IT Programme and Project Delivery	IT Operations and Service Delivery	Negative Example Scenarios	Positive Example Scenarios
0601	Information (data breach: damage, leakage and access)	S		P	Hardware components are damaged, leading to (partial) destruction of data by internal staff.	Backup procedures, aligned to the business criticality of the data, are established, ensuring key business data is always retained at a second location.
0602		S	S	P	The database is corrupted, leading to inaccessible data.	
0603		S	S	P	Portable media containing sensitive data (CD, USB drives, portable disks, etc.) is lost/disclosed.	Portable media are appropriately secured and encrypted to ensure protection of data.
0604		S	S	P	Sensitive data is lost/disclosed through logical attacks.	Sensitive data residing in the enterprise premises are protected appropriately behind firewalls and through continuous network monitoring.
0605		S	S	P	Backup media is lost or backups are not checked for effectiveness.	
0606		P	S	P	Sensitive information is accidentally disclosed due to failure to follow information handling guidelines.	Employees are encouraged continuously to be ambassadors of the enterprise culture, ethics and good behaviours, including practices around information handling.
0607		P	S	P	Data (accounting, security-related data, sales figures, etc.) are modified intentionally.	The 4-eye principle is applied for specific data inputs/modifications to create a peer review and decrease the stimulus for intentional modification.
0608		P	S	P	Sensitive information is disclosed through email or social media.	Employees are encouraged continuously to be ambassadors of the enterprise culture, ethics and good behaviours, including practices involving distribution of information through email and social media.
0609		P	S	P	Sensitive information is discovered due to inefficient retaining/archiving/disposing of information.	The data retention policy is updated regularly and strict compliancy is endorsed for all employees.
0610		P	S	P	IP is lost and/or competitive information is leaked due to key team members leaving the enterprise.	IP clauses are incorporated in every contract, allowing the enterprise to fully reap the benefits of all IP created in the enterprise.
0611		P	S	P	The enterprise has an overflow of data and cannot deduct the business relevant information from the data (e.g., big data problem).	The enterprise has an effective process in place to process the data it has into business relevant information and use that information to create business value.
0701	Architecture (architectural vision and design)	P	P	P	The enterprise architecture is complex and inflexible, obstructing further evolution and expansion leading to missed business opportunities.	Modern and flexible architecture supports business agility/innovation.
0702		P	S	P	The enterprise architecture is not fit for purpose and not supporting the business priorities.	
0703		P	S	S	There is a failure to adopt and exploit new infrastructure in a timely manner.	
0704		P	S	S	There is a failure to adopt and exploit new software (functionality, optimisation, etc.) in a timely manner.	

Figure 38—Example Risk Scenarios *(cont.)*

		Risk Type			Example Scenarios	
					Figure 38—Example Risk Scenarios *(cont.)*	
Ref.	Risk Scenario Category	IT Benefit/Value Enablement	IT Programme and Project Delivery	IT Operations and Service Delivery	Negative Example Scenarios	Positive Example Scenarios
0801	Infrastructure (hardware, operating system and controlling technology) (selection/ implementation, operations and decommissioning)	P	S	P	New (innovative) infrastructure is installed and as a result systems become unstable leading to operational incidents, e.g., Bring your own device (BYOD) programme.	Appropriate testing is conducted before setting infrastructure into the production environment to ensure the availability and proper functioning of the entire system.
0802		P	S	P	The systems cannot handle transaction volumes when user volumes increase.	
0803		P	S	P	The systems cannot handle system load when new applications or initiatives are deployed.	
0804		P	S	P	Intermittently, there are failures of utilities (telecom, electricity).	Second line utilities are foreseen and stand by 24/7 to support the continuous execution of business critical transactions.
0805		P	S	P	The IT in use is obsolete and cannot satisfy new business requirements (networking, security, database, storage, etc.).	IT is an innovator, ensuring a two-way interaction between business and IT.
0806				P	Hardware fails due to overheating.	
0901	Software	P		S	There is an inability to use the software to realise desired outcomes (e.g., failure to make required business model or organisational changes).	The software in use stimulates the generation of new ideas.
0902		P		S	Immature software (early adopters, bugs, etc.) is implemented.	
0903		P		S	The wrong software (cost, performance, features, compatibility, etc.) is selected for implementation.	Upfront analysis is performed and a business case is made up to ensure the adequate selection of software.
0904		P		S	There are operational glitches when new software is made operational.	User adapted training and user acceptance testing is performed before the go-live decision to ensure the smooth transition to new software and that generation of business value continues.
0905		P		S	Users cannot use and exploit new application software.	
0906		P		S	Intentional modification of software leading to wrong data or fraudulent actions.	The 4-eye principle is applied for specific data inputs / modifications to create a peer review and decrease the stimulus for fraudulent actions or simply unexpected results.
0907		P		S	Unintentional modification of software leads to unexpected results.	
0908		P		S	Unintentional configuration and change management errors occur.	Enterprisewide configuration management decreases resolution time for incident and problem management.
0909		P		S	Regular software malfunctioning of critical application software occurs.	Appropriate testing is conducted before the go-live decision to ensure the availability and proper functioning of the software.
0910		P		S	Intermittent software problems with important system software occur.	
0911		P		S	Application software is obsolete (e.g., old technology, poorly documented, expensive to maintain, difficult to extend, not integrated in current architecture)	IT is an innovator, ensuring a two-way interaction between business and IT.
0912		P		S	There is an inability to revert back to former versions in case of operational issues with the new version.	Backup and restore points are established in accordance with business criticality of software to ensure roll-back procedures.

		Risk Type			Example Scenarios	
Ref.	Risk Scenario Category	IT Benefit/Value Enablement	IT Programme and Project Delivery	IT Operations and Service Delivery	Negative Example Scenarios	Positive Example Scenarios
1001	Business ownership of IT	P	P	S	Business does not assume accountability over those IT areas it should, e.g., functional requirements, development priorities, assessing opportunities through new technologies.	Business assumes appropriate accountability over IT and co-determines the strategy of IT, especially application portfolio.
1002		P	S	S	There is extensive dependency and use of end-user computing and *ad hoc* solutions for important information needs, leading to security deficiencies, inaccurate data or increasing costs/inefficient use of resources.	
1003		P	S	S	Cost and ineffectiveness is related to IT related purchases outside of the procurement process.	A business case is always made up to ensure optimal cost and effective purchasing of software.
1004				P	Inadequate requirements lead to ineffective service level agreements (SLAs).	
1101	Supplier selection/ performance, contractual compliance, termination of service and transfer		S	P	There is a lack of supplier due diligence regarding financial viability, delivery capability and sustainability of supplier's service.	Third party acts as strategic partner.
1102			S	P	Unreasonable terms of business are accepted from IT suppliers.	
1103			S	P	Support and services delivered by vendors are inadequate and not in line with the SLA.	Appropriate key performance indicators (KPIs), linked to rewards and penalties, ensure adequate service delivery and support.
1104			S	P	Outsourcer performance is inadequate in a large-scale long-term outsourcing arrangement.	
1105			S	P	There is non-compliance with software licence agreements (use and/or distribution of unlicenced software, etc.).	Contractual arrangements are agreed on concerning the use of third-party software and proprietary software.
1106			S	P	There is an inability to transfer to alternative suppliers due to overreliance on current supplier.	A phase-out and knowledge transfer clause is added to the contract with the supplier, enforcing it to do a handover with new suppliers. A mix of internal and external employees is set up for each process, avoiding full knowledge of the process only residing with external employees.
1107			S	P	Cloud services are purchased by the business without the consultation/involvement of IT, resulting in inability to integrate the service with in-house services.	
1201	Regulatory compliance	P	S	S	There is non-compliance with regulations, e.g., privacy, accounting, manufacturing.	Full compliance with regulations is exploited towards clients to generate extra business value.
1202		P	S	S	Unawareness of potential regulatory changes have an impact on the operational IT environment.	The enterprise sets up a legal and compliance department to follow up on regulatory changes and to ensure the continuation of business value generation.
1203		P	S	S	The regulator prevents cross-border dataflow due to insufficient controls.	

Figure 38—Example Risk Scenarios (cont.)

		Risk Type			Example Scenarios	
Ref.	Risk Scenario Category	IT Benefit/Value Enablement	IT Programme and Project Delivery	IT Operations and Service Delivery	Negative Example Scenarios	Positive Example Scenarios
1301	Geopolitical			P	There is no access due to disruptive incident in other premises.	Clear compliance with national policies and support of local initiatives ensures support by local government and generation of business value.
1302				P	Government interference and national policies limit service capability.	
1303				P	Targeted action against the enterprise results in destruction of infrastructure.	
1401	Infrastructure theft or destruction	S	S	P	There is a theft of a device with sensitive data.	Key infrastructure components are monitored 24/7 for performance, availability, etc. Alarm bells are raised in case of irregularities and acted on immediately.
1402		S	S	P	There is a theft of a substantial number of development servers.	
1403		S	S	P	Destruction of the data centre (sabotage, etc.) occurs.	Data centre is appropriately secured, only allowing access to authorised IT staff.
1404		S	S	P	There is accidental destruction of individual devices.	
1501	Malware	S		P	There is an intrusion of malware on critical operational servers.	IT infrastructure will be appropriately protected behind firewalls and through continuous monitoring of the network to ensure the execution of day-to-day activities.
1502		S		P	Regularly, there is infection of laptops with malware.	
1503		S		P	A disgruntled employee implements a time bomb that leads to data loss.	
1504		S		P	Company data are stolen through unauthorised access gained by a phishing attack.	
1601	Logical attacks	S		P	Unauthorised users try to break into systems.	
1602		S		P	There is a service interruption due to denial-of-service attack.	
1603		S		P	The web site is defaced.	
1604		S		P	Industrial espionage takes place.	
1605		S		P	There is a virus attack.	
1606		S		P	Hacktivism takes place.	
1701	Industrial action	S	S	P	Facilities and building are not accessible because of a labour union strike.	A business continuity plan foresees action to be taken to always ensure the execution of business critical tasks in case the building is not accessible anymore.
1702		S	S	P	Key staff is not available through industrial action (e.g., transportation strike).	A flexible work policy, allowing employees to work from another location than the office building simulates freedom and creates a positive work atmosphere.
1703		S	S	P	A third party is not able to provide services because of strike.	
1704		S	S	P	There is no access to capital caused by a strike of the banking industry.	

Figure 38—Example Risk Scenarios (cont.)

Figure 38—Example Risk Scenarios *(cont.)*					
		Risk Type			Example Scenarios	
Ref.	Risk Scenario Category	IT Benefit/Value Enablement	IT Programme and Project Delivery	IT Operations and Service Delivery	Negative Example Scenarios	Positive Example Scenarios
1801	Environmental	S	S	P	The equipment used is not environmentally friendly (e.g., power consumption, packaging).	Being awarded for environmental friendliness creates positive media attention, attracts new customers and employees, and ensures value creation.
1901	Acts of nature	S	S	P	There is an earthquake.	
1902		S	S	P	There is a tsunami.	
1903		S	S	P	There are major storms and tropical cyclones.	
1904		S	S	P	There is a major wildfire.	
1905		S	S	P	There is flooding.	
1906		S	S	P	The water table is rising.	
2001	Innovation	P	S	S	New and important technology trends are not identified.	Innovation and trend watch are endorsed and encouraged, ensuring new technology (trends) are timely assessed for business impact and adopted if required.
2002		P		S	There is a failure to adopt and exploit new software (functionality, optimisation, etc.) in a timely manner.	Innovation and trend watch are endorsed and encouraged, ensuring new technology (trends) are timely assessed for business impact and adopted if required.
2003		P		S	New and important software trends are not identified (consumerisation of IT).	

Appendix D provides a set of examples of how COBIT 5 enablers can help to mitigate the risk scenarios in **figure 38**. Other IT management frameworks, such as ITIL and ISO/IEC 27001/2, can also be used to that purpose, but no detailed links/mappings are included in this document.

CHAPTER 4
RISK AGGREGATION

4.1 Why Risk Aggregation?

IT risk management can only reach its full potential if risk is managed throughout the entire enterprise. It is less valuable when only a partial view on risk is obtained. Partial has two aspects in this context:
• Only part of potential risk items are considered during risk analysis and risk management.
• Only part of the enterprise is within scope of risk management, i.e., not the entire enterprise is considered.

Every enterprise needs an end-to-end (business activity) aggregated view of risk, beyond the technical issues, to prevent a false sense of security or a false sense of urgency. An aggregated risk view allows proper review of risk appetite and tolerance, instead of only having silo views of individual or partial risk items, e.g., a change management problem on an ERP system could have far reaching consequences across multiple business lines, countries, partners and customers. The executive management needs to see the aggregated impact of this risk to the entire enterprise, not just see it is a risk on one server in one data centre in one location.

In practice, some obstacles prevent effectively obtaining a consistent, aggregated and realistic view on actual exposure at the enterprise level:
• Lack of consistent and clear terminology across the enterprise
• Complex enterprises with different (sub-)cultures exist, making it difficult to contain and define the different entities where risk needs to be described, and making it difficult to obtain coherent, reliable and consistent data on risk, even the absolute minimum requirement of a high-level risk assessment.
• Presence of qualitative data (and absence of quantitative data) in most cases, with limited reliance on the reliability of the reported risk levels, or with different and/or incompatible scales used for assessing impact and frequency
• Unknown dependencies between reported risk can hide bigger risk, e.g., different entities all reporting same, medium-level risk, which can actually turn out to be a major risk for the entire enterprise when occurring.
• The use of ordinal scales for expressing risk in different categories, and the mathematical difficulties or dangers of using these numbers to do any sort of calculation
• In complex enterprises, a particular risk at entity level may be important for the entity itself, but for various reasons (e.g., size, enterprise strategy) may be less important at enterprise level. This scaling of risk needs to be well understood when aggregating risk information.
• Different stakeholders (such as operational risk groups, internal audit, technology risk management, governance, business process improvement [BPI], project management office [PMO], enterprise architecture [EA], quality control [QC]) of large and complex enterprises use different methodologies/frameworks to understand, measure and respond to risk. This prevents an effective aggregation of risk.
• Immaturity of the enterprise in terms of process management, leading to failure to measure process performance and process outcomes, hence inability to have an accurate view on risk factors
• Failure to understand gaps in both the ability to detect an event occurring and to respond to it

4.2 Approach Towards Risk Aggregation

There are different ways to perform risk aggregation. Some guidance in this respect follows:
• Ensure there is a uniform, consistent, agreed-on and communicated method for assessing frequency and impact of risk scenarios. The same method should be used to present aggregated risk. Using a consistent taxonomy for describing risk allows aggregating of and reporting on varying types of risk, e.g., value creation related risk, project delivery risk and operational risk, because they are all expressed in terms of business impact using the same metrics.
• Be cautious with the mathematics, and only aggregate data and numbers that are meaningful. Do not aggregate data of different nature, e.g., on status of controls or operational IT metrics. Although separately these are good risk indicators, they are meaningless when they are not associated with an ultimate business impact. For example, if certain controls are not fully effective or are badly designed, this constitutes no risk in itself. There is an issue only when the risk scenarios that rely on these controls are unacceptable because of the failing controls. Hence, the information on failing controls is not a reliable metric on its own.

- Focus on real risk for business activities and the most important indicators thereof, and avoid focusing on adding up things that are easily measurable but less relevant. Reporting firewall attacks may be easy to measure, but if up-to-date security measures are in place, these attacks, although probably very frequent, carry little business impact.
- Do not aggregate risk information in such a way that it hides actionable detail. This may occur because of the organisational reporting 'level of responsibility' issues that must be addressed by a certain organisational layer and must be visible to that layer, but may be aggregated and hidden from the next-higher level of authority because no immediate action is required by that level. The root cause of risk must be visible to those responsible for managing it. Attention must be given to the aggregation algorithm that will be used.
- Aggregation is possible in multiple dimensions, e.g., organisational units, types of risk items, business processes. The benefit of aggregation in business processes is that it reveals weak links to achieving successful business outcomes. Sometimes multiple views (using a combination of several dimensions) may be needed to satisfy risk management and business needs.
- Aggregate risk at the enterprise level, where risk can be considered in combination with all other risk the enterprise needs to manage (integration with ERM). Take into account the organisational structure (geographical split, business units, etc.) to set up a meaningful cascade of risk aggregation without losing sight of important specific risk items.
- Take into account dependencies at different levels:
  - The dependency between event and ultimate business impact needs to be understood. For example, if a server goes down, which business process might suffer, and how does this translate into financial impact, impact for the customer, etc.? This is part of the initial risk analysis process.
  - Dependencies between events make aggregation more than a mathematical addition. Events can amplify each other. For example:
    - One data centre down could be acceptable, but a second data centre down at the same time might be a catastrophe.
    - A security incident, followed by an error during security software emergency upgrade (because of inadequate change management and configuration management procedures) leads to extended recovery times for critical services affected.
    - A project developing a new IT architecture, including data models, infrastructure, etc., is significantly delayed, thus delaying several new application development projects which would normally rely on the timely completion of the architecture project.

Figure 39 shows one possible, simple, approach for aggregating risk. This approach is only valid when risk items are disjointed (independent) between entities. When risk items are shared or connected, this approach is not valid and may lead to underestimation of actual risk. In the **figure 40** example:
- Two entities create, after due risk analysis, their own risk maps. Notice that the entity on the right has more severe risk compared to the entity on the left.
- The risk scenarios on the maps are brought together on one aggregated map. This approach is only valid when all entities use the same metrics and scales in their risk maps.
- The aggregated picture shows an even spread of risk throughout the enterprise, allowing proper management response to be defined.

The method to aggregate risk shown in **figure 39** was by simple adding; other methods may exist, e.g., each entity only shows top ten risk items. As mentioned previously, the aggregation method shown must still allow sufficient sensitivity, i.e., no major risk or root causes should remain hidden from the appropriate decision makers.

The aggregation approach described previously is only valid when risk items in different entities are independent, i.e., when they are not shared and when they do not influence each other. Therefore, a key activity in risk aggregation is to analyse risk analysis results from different entities, and verify whether such dependencies exist. When they do exist, adapt the aggregated risk map accordingly.

If for example, all units were using the same data centre or power grid, and when this data centre or power grid became unavailable, then the entire enterprise is affected at once. This may be assessed differently, i.e., more seriously, compared to one entity going down for a brief while.

When discussing cascading risk, the magnitude of a joint failure might be increased, but generally, the frequencies of joint failures are less. In other words, the probability of two or more elements failing at the same time is invariably less than the probability of either one failing.

The aggregated risk map may then look as shown in **figure 40**.

Figure 39—Aggregation of Risk Maps—Disjoint Risk

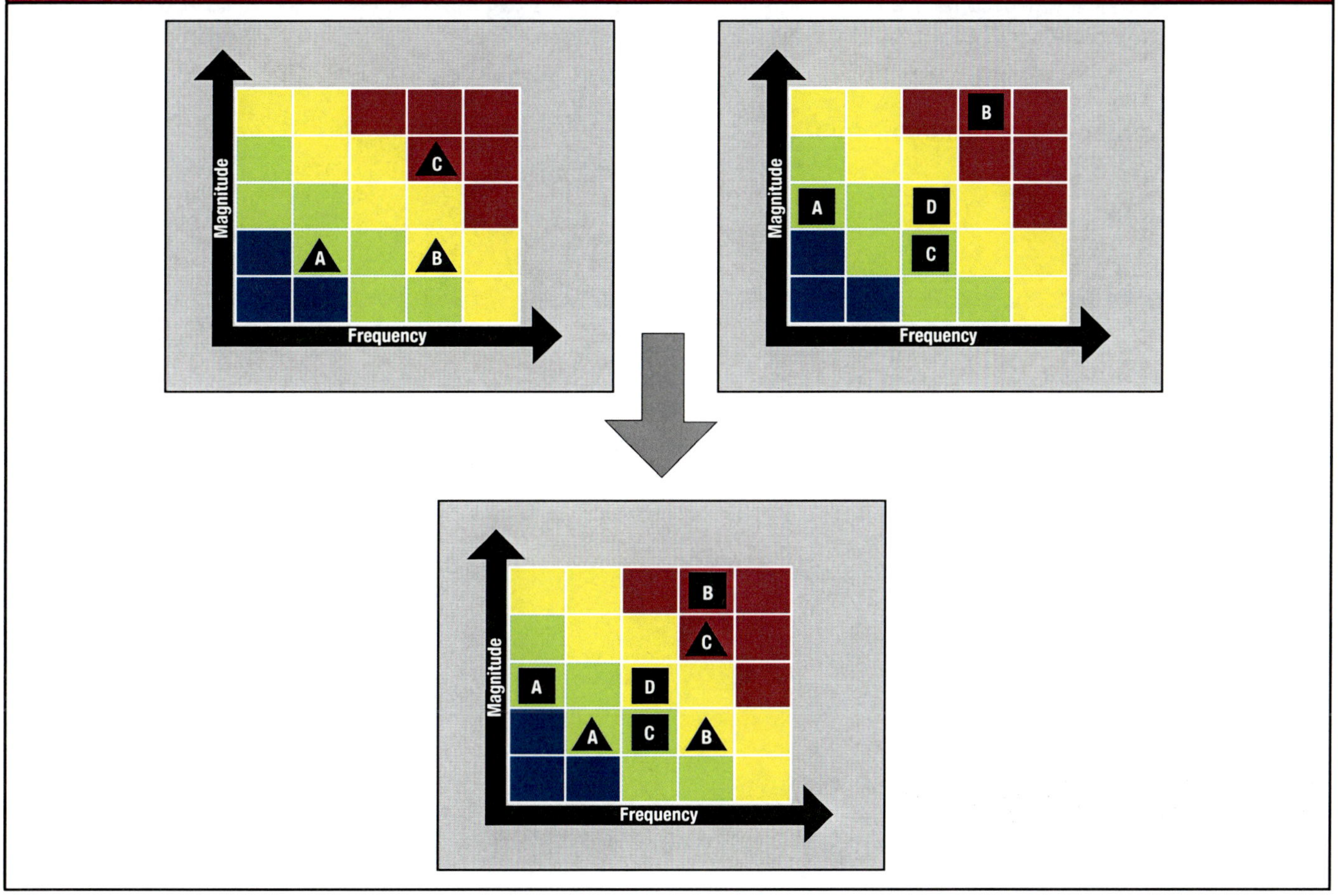

Figure 40—Aggregation of Risk Maps—Shared Risk

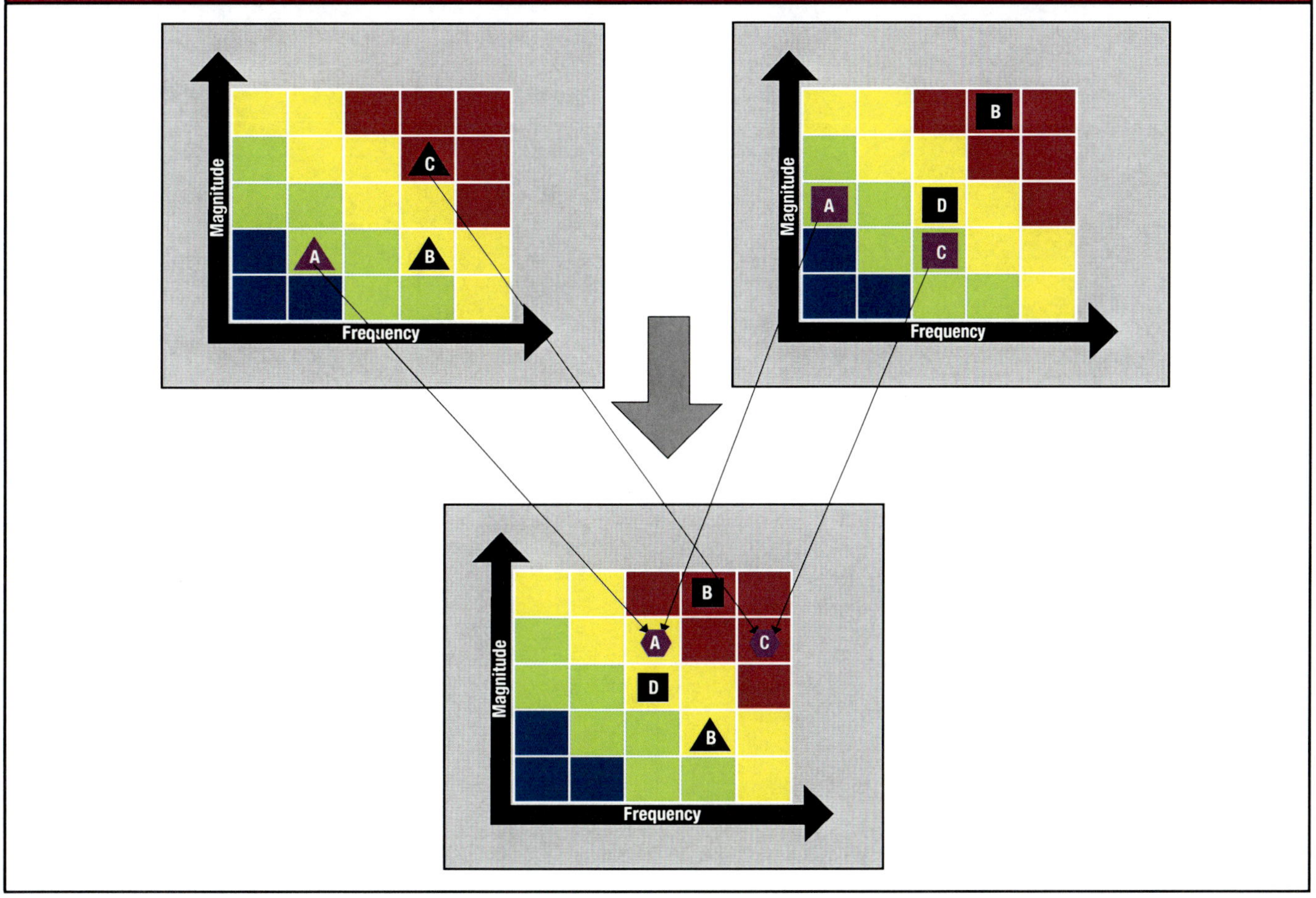

One benefit of aggregation, and certainly in the case of dependencies, is that the risk for the overall enterprise becomes highly visible and funds may become available to define an enterprisewide response to the risk. Whereas at the entity level, such a response would not have been feasible or justifiable. Aggregation has thus allowed definition and implementation of cost-efficient response to current risk, and reduction of residual risk within the defined risk appetite levels.

Aggregated risk maps can be part of the risk profile of the enterprise, which itself is part of risk reporting.

CHAPTER 5
RISK RESPONSE

5.1 Definitions

The concepts in **figure 41** are frequently used in risk management and in this document.

Figure 41—Defined Risk Terms	
Term	**Definition**
Risk appetite	The amount of risk, on a broad level, an entity is willing to accept in pursuit of its mission
Risk tolerance	The acceptable level of variation that management is willing to allow for any particular risk as it pursues objectives
Risk capacity	The objective amount of loss an enterprise can tolerate without risking its continued existence. As such, it differs from risk appetite, which is more a board/management decision on how much risk is desirable.

These concepts are further illustrated in appendix B.5.6

5.2 Risk Response Workflow and Risk Response Options

The purpose of defining a risk response is to bring risk in line with the defined risk appetite for the enterprise. In other words, a response needs to be defined such that as much future residual risk (current risk with the risk response defined and implemented) as possible (usually depending on budgets available) falls within risk tolerance limits. The full risk response workflow is depicted in **figure 42**. Corresponding guidance with regard to these activities can be found in COBIT 5 processes EDM03 and APO12, more specifically in practices EDM03.02 and APO12.02.

This risk response evaluation is not a one-time effort; rather, it is part of the risk management process cycle. When risk analysis of all identified risk scenarios, after weighing risk vs. potential return, has shown that risk is not aligned with the defined risk appetite and tolerance levels, a response is required. This response can be any of the four possible responses explained in the following subsections.

Risk Avoidance
Avoidance means exiting the activities or conditions that give rise to risk. Risk avoidance applies when no other risk response is adequate. This is the case when:
• There is no other cost-effective response that can succeed in reducing the frequency and impact below the defined thresholds for risk appetite.
• The risk cannot be shared or transferred.
• The exposure level is deemed unacceptable by management.

Some IT-related examples of risk avoidance may include:
• Relocating a data centre away from a region with significant natural hazards
• Declining to engage in a very large project when the business case shows a notable risk of failure
• Declining to engage in a project that would build on obsolete and convoluted systems because there is no acceptable degree of confidence that the project will deliver anything workable
• Deciding not to use a certain technology or software package because it would prevent future expansion

Figure 42—Risk Response Workflow

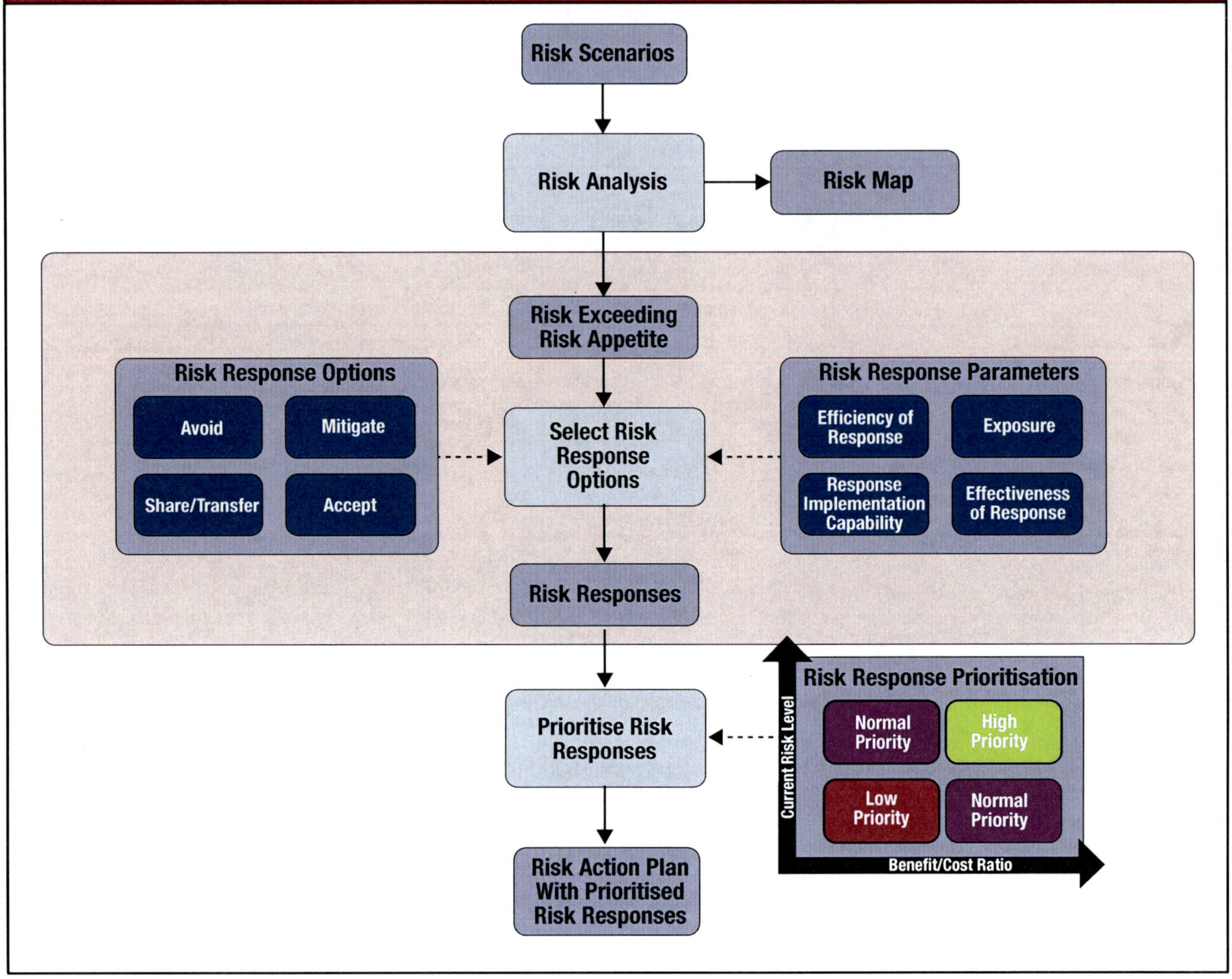

Risk Scenarios

Risk Analysis → Risk Map

Risk Exceeding Risk Appetite

Risk Response Options
- Avoid
- Mitigate
- Share/Transfer
- Accept

Select Risk Response Options

Risk Response Parameters
- Efficiency of Response
- Exposure
- Response Implementation Capability
- Effectiveness of Response

Risk Responses

Prioritise Risk Responses

Risk Response Prioritisation

Current Risk Level / Benefit/Cost Ratio
- Normal Priority
- High Priority
- Low Priority
- Normal Priority

Risk Action Plan With Prioritised Risk Responses

Risk Acceptance

Acceptance means that exposure to loss is recognised but no action is taken relative to a particular risk, and loss is accepted when/if it occurs. This is different from being unaware of risk. Accepting risk assumes that the risk is known, i.e., an informed decision has been made by management to accept it as such (e.g., when cost of remediation outweighs the risk).

If an enterprise adopts a risk acceptance stance, it should carefully consider who can accept the risk—even more so with IT risk. IT risk should be accepted only by business management (and business process owners), in collaboration with and supported by IT, and acceptance should be communicated (i.e., documented) to senior management and the board (Refer to EDM03.02 detailed activities 5.3 and 5.4). An enterprise should also consider establishing authorised risk acceptance levels, setting out whom within the enterprise is authorised to accept different levels of risk. This will help ensure that risk is accepted at the right level within the enterprise.

Some examples of risk acceptance are:
- There may be risk that a certain project will not deliver the required business functionality by the planned delivery date. Management may decide to accept the risk and proceed with the project.
- If a particular risk is assessed to be extremely rare but very important (catastrophic) and approaches to reduce it are prohibitive, management may decide to accept it.

Self-insurance is another form of risk acceptance, although this manages only magnitude of the loss and has no impact on frequency.

Risk Sharing/Transfer

Sharing means reducing risk frequency or impact by transferring or otherwise sharing a portion of the risk. Common techniques include insurance and outsourcing. Examples include taking out insurance coverage for IT-related incidents, outsourcing part of the IT activities, or sharing IT project risk with the provider through fixed-price arrangements or shared-investment arrangements. In both a physical and legal sense, these techniques do not relieve an enterprise of the risk ownership, but can involve the skills of another party in managing the risk and reduce the financial consequence if an adverse event occurs. Also from a reputation point of view, risk transfer or sharing does not transfer ownership or accountability over risk.

Some IT-related examples of risk sharing or transfer are:
• A large enterprise identified and assessed the risk of fire to its infrastructure across diverse geographic regions and assessed the cost of sharing the impact of its risk through insurance coverage. It concluded that, because of the location of its sites, the incremental cost of insurance and related deductibles was not prohibitive, and insurance coverage was taken.
• In a major IT-related investment, project risk may be shared by outsourcing the development to an outsourcer for a fixed price on a risk/reward basis.
• Some enterprises outsource some or all of their IT function to hosting enterprises and contractually share a portion of the risk.
• Where application hosting is outsourced, the enterprise always remains accountable for protecting client privacy, but if the outsourcer is negligent and a breach occurs, risk (financial impact) might at least be shared with the outsourcer.

Other techniques contributing to risk sharing include:
• Large enterprises with multiple legal entities, where IT risk can be transferred to other divisions within the enterprise (reinsurance is a common example)
• SSAE16 certification, which allows a service organisation to transfer a portion of a risk back to the client through the client considerations section of the report

Risk Mitigation

Risk mitigation means that mitigating action is taken to reduce the frequency and/or impact of a risk. The most common ways of mitigating risk include:
• Strengthening overall IT risk management practices, i.e., implement sufficiently mature IT risk management processes as defined by the COBIT 5 framework
• Introducing a number of control measures intended to reduce either frequency of an adverse event happening and/or the business impact of an event, should it happen. Controls are, in the context of risk management, employed to mitigate a risk, e.g., the policies, procedures and practices, structures, information flows, etc. The COBIT 5 set of interconnected enablers provides a comprehensive set of controls that can be implemented. It is possible to identify, for any given risk scenario that would exceed risk appetite, a set of COBIT 5 enablers (processes, organisational structures, behaviours, etc.) that can mitigate the risk scenario. For a comprehensive list of controls (expressed as COBIT 5 enablers) that can mitigate risk (list of example generic risk scenarios as defined in chapter 3), refer to appendix D.
• Mitigation of risk is possible by other means or methods, e.g., there are well-known IT management frameworks and standards able to assist.

5.3 Risk Response Selection and Prioritisation

The previous section discussed available risk response options. In this section, the selection of an appropriate response, i.e., given the risk at hand, how to respond and how to choose between the available response options, is discussed briefly. As illustrated in **figure 43**, the following parameters need to be taken into account in this process:
• Efficiency of the response, i.e., the relative benefits promised by the response in comparison to:
  – Other investments (investing in risk response measures always competes with other investments)
  – Other responses (one response may address several risk items while another may not)

  This also takes into account the cost of the response, e.g., in the case of risk transfer, the cost of the insurance premium; in the case of risk mitigation, the cost (capital expense, salaries, consulting) to implement control measures.
• Exposure, i.e., the importance of the risk addressed by the response, represented by its position on the risk map (which reflects combined frequency and impact levels)
• The enterprise's capability to implement the response. When the enterprise is mature in its risk management processes, more sophisticated responses can be implemented; when the enterprise is rather immature, it may be better to start with some basic responses.
• Effectiveness of the response, i.e., the extent to which the response will reduce the impact and magnitude of the risk

Figure 43—Risk Response Prioritisation Workflow

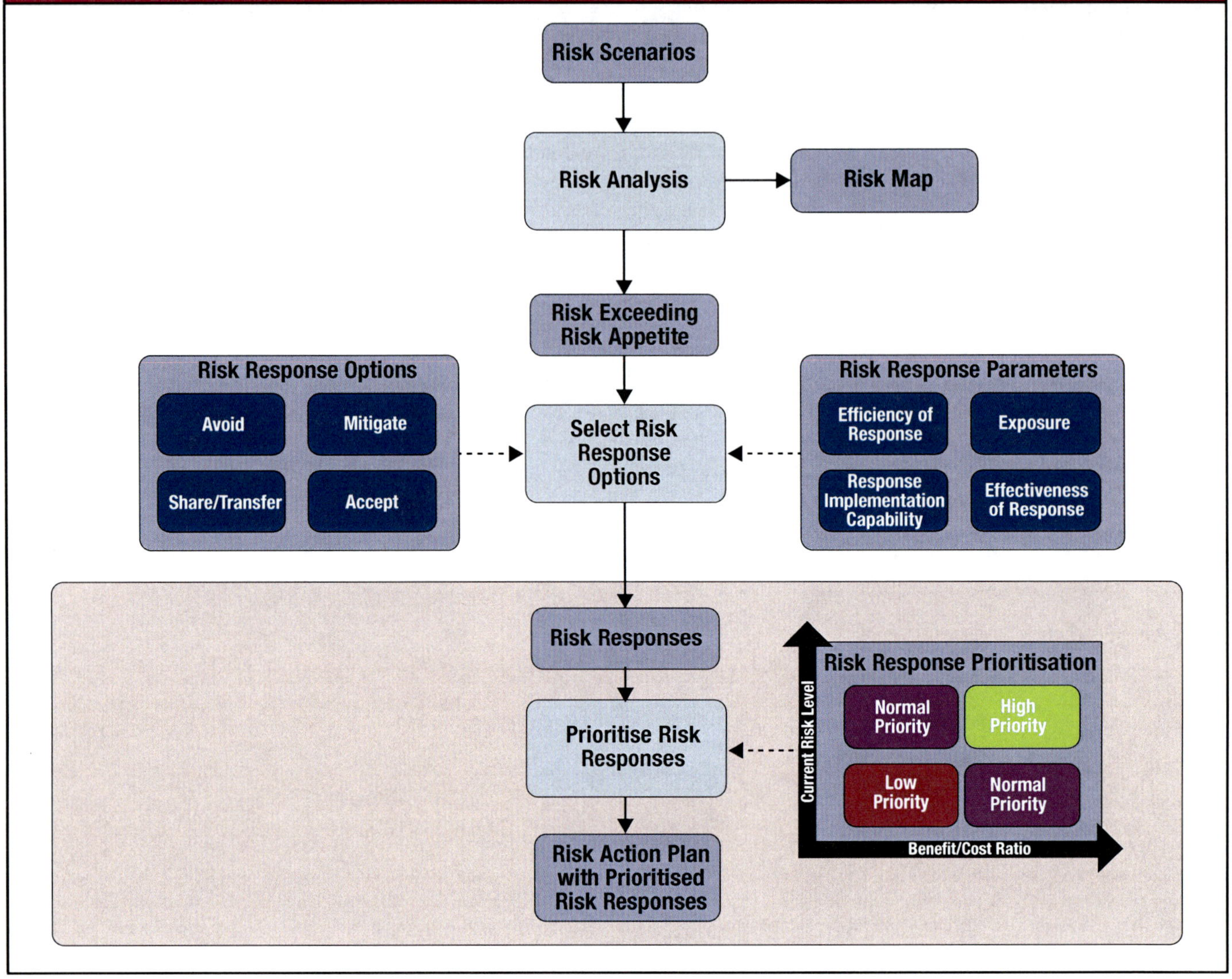

The aggregated required effort for the mitigation responses, i.e., the collection of controls that need to be implemented or strengthened, may exceed available resources. As for all investments, a business case needs to be prepared and prioritisation is required, provided there is no urgency required. Using the same criteria as for risk response selection, risk responses can be placed in a quadrant offering **three possible priorities**, as shown in **figure 43**.
- **High priority**—Very cost-efficient and effective responses on high risk
- **Normal priority**—More expensive or difficult responses to high risk or efficient and effective responses on lower risk items, both requiring careful analysis and management decision on investments
- **Low priority**—Responses to lower risk that may prove costly or may not demonstrate an effective cost/benefit ratio

Some examples of prioritisation include:
- A risk has been identified that the enterprise's IT architecture is so complex that within a few years' time extending capacity will become difficult and maintenance of software will become very expensive. The identified responses consisted of strengthening the COBIT APO03 *Manage enterprise architecture* process, as well as starting a large project to overhaul the complete architecture. Given the cost of this project, this response was categorised as 'normal'.
- A risk of non-compliance with regulations was identified because of a number of missing relatively simple IT procedures. The response consisted of creating the missing IT procedures and implementing them. This was classified as 'high'.

5.4 Guidance on Risk Response Selection and Prioritisation

The organisational structures in *COBIT 5 for Risk* include RACI (responsible, accountable, consulted, informed) charts for all IT risk management process activities, including risk response definition and prioritisation. All major stakeholders should be involved in those decisions, i.e., senior management, business management, IT and risk management.

Based on the outcomes of risk analysis and the experience gained during risk response definition and prioritisation, the enterprise may also decide on more fundamental changes in its position against risk, for example:
• Review the risk appetite thresholds or temporarily increase or decrease risk appetite levels
• Increase (or decrease) resources available for executing risk response
• Accept risk that normally would exceed risk appetite thresholds

In the evaluation and design of risk responses, enterprises should always look for a balanced set of responses, i.e., a combination of awareness/education, process/governance and automation.

Page intentionally left blank

SECTION 3. HOW THIS PUBLICATION ALIGNS WITH OTHER STANDARDS

COBIT 5 for Risk—much like COBIT 5 itself—is an umbrella framework for the governance and management of risk. To better understand this umbrella position, this section contains a positioning of *COBIT 5 for Risk* against the following IT-risk-related standards.
• ISO 31000
• ISO/IEC 27005
• COSO ERM

CHAPTER 1
COBIT 5 FOR RISK AND ISO 31000

1.1 ISO 31000:2009 Risk Management Principles and Guidelines

This standard contains three major clauses:
• Chapter 4. Principles for managing risk
• Chapter 5. Risk management framework
• Chapter 6. Processes for managing risk

Overall, it can be noted that the process as defined in ISO 31000 is fully covered by the different processes and practices of the *COBIT 5 for Risk* process model. *COBIT 5 for Risk*, however, provides more extensive guidance and includes areas not covered by ISO 31000, such as risk governance.

Principles for Managing Risk
Figure 44 contains the 11 risk management principles ISO 31000 has defined and the extent to which *COBIT 5 for Risk* covers them.

Figure 44—ISO 31000 Risk Management Principles Covered by *COBIT 5 for Risk*	
ISO 31000 Principles	***COBIT 5 for Risk* Coverage**
Risk management creates and protects value.	• Principle 1: Managing stakeholder needs. • Each enabler provides value by the achievement of its goals.
Risk management is an integral part of organisational processes.	• Principle 2: Covering the enterprise end-to-end. • Features a process model that promotes integration into enterprise risk management (ERM) processes and operational processes (APO12 *Manage risk*).
Risk management is part of decision making.	• Dedicates an entire process (EDM03 *Ensure risk optimisation*) to risk-aware business decisions. • The capability models illustrate how enterprise decisions improve based on increasing and appropriate stakeholder involvement and improving quality and availability of risk analysis results. • The process model provides a full set of RACI charts, indicating how risk management accountabilities and responsibilities can be assigned throughout the enterprise.
Risk management explicitly addresses uncertainty.	• Recommends management practices that estimate IT risk-based on scenarios of varying frequency and impact. Scenarios enable considering different factors enabling or creating uncertainty.
Risk management is systemic, structured and timely.	• Enabler: Processes • Enabler: Principles, Policies and Frameworks • The enabler dimension life cycle is a systemic and structured way of managing risk. • The enabler dimension good practices provide management practices that consist of process practices, activities and detailed activities.
Risk management is based on the best available information.	• Enabler: Information
Risk management is tailored.	• Principle 1: Managing stakeholder needs • Describes processes (EDM03, MEA01, MEA02 and MEA03) that accommodate the enterprise's unique performance needs and external requirements.

Figure 44—ISO 31000 Risk Management Principles Covered by *COBIT 5 for Risk (cont.)*	
ISO 31000 Principles	***COBIT 5 for Risk* Coverage**
Risk management takes human and cultural factors into account.	• Enabler: Culture Ethics and Behaviour • Provides example scenarios that cover human factors.
Risk management is transparent and inclusive.	• Principle 4: Promotes fair and open communication of IT risk. • Supporting process EDM05 ensures stakeholder transparency. • Process APO12 promotes transparency and inclusiveness of risk management. • Recommends transparency as an enterprisewide behaviour.
Risk management is dynamic, iterative and responsive to change.	• Principle 1: Managing stakeholder needs. • Dedicates an entire process (APO12) to maintaining the IT risk profile to help IT risk management activities stay in sync with organisational changes. • Describes how to use the seven enablers to respond to risk. • Each enabler in has a life cycle. The phases evaluate/monitor and update/dispose ensure a dynamic, iterative and responsive character.
Risk management facilitates continual improvement of the organisation.	• Includes management practices and information flows supporting process improvements based on lessons learned from risk events, policy exceptions, and data on risk-aware culture change. • The process model includes goals and metrics (MEA02) that can be used to measure performance.

Risk Management Framework

ISO 31000 defines a five-block risk management framework. **Figure 45** contains the five blocks (labelled A through E) and explains how *COBIT 5 for Risk* addresses each of these components.

Figure 45—ISO 31000 Risk Management Framework Covered by *COBIT 5 for Risk*	
ISO 31000 Framework Components	***COBIT 5 for Risk* Coverage**
A. Mandate and Commitment	• Includes practices to align IT risk management objectives and performance indicators with those of ERM. • Defines IT risk management roles and suggests the assignment of responsibility and accountability for key activities. • The enabler models include stakeholder specific content, such as specific information to be communicated, roles and responsibilities. • Enabler: Principles, Policies and Frameworks; details core and other risk policies • Includes a dedicated process for the evaluation, monitoring, and assessment of compliance with external requirements (MEA03)
B. Framework Design for Managing Risk	
1. Understanding the organisation and its context	• Enabler: Organisational Structures • Includes practices to collect data on the operating environment. • Includes management practices to understand the internal context and determine where/how enterprise processes rely on IT for success.
2. Risk management policy	• Enabler: Principles, Policies and Frameworks • Includes management practices to codify risk appetite and tolerance into policy and align existing IT-related policies with approved risk tolerance. • Includes escalation paths to deal with conflicting situations related to application of policy.
3. Accountability	• Enabler: People, Skills and Competencies • Defines IT risk management roles and suggests the assignment of responsibility and accountability for key activities to these roles. • Includes management practices to establish and maintain accountability for IT risk management.
4. Integration into organisational processes	• Enabler: Processes. • Principle 2: Covering the enterprise end-to-end. • Dedicates an entire process to integrate with ERM. • Includes detailed linkages to COBIT 5 that model a wide range of IT and business processes.
5. Resources	• Includes management practices to provide adequate resources for IT risk management (APO07).
6. Establishing internal communication and reporting mechanisms	• Enabler: Information • Enabler: Services, Infrastructure and Applications • Includes management practices to encourage effective communication of IT risk (EDM03). • Process model includes specific information to be communicated between the key management practices. • The framework introduction features a section on risk communication with suggested information flows between different stakeholders.
7. Establishing external communication and reporting mechanisms	• Enabler: Information • Enabler: Services, Infrastructure and Applications • Includes management practices to communicate and report ongoing risk management activities and communicate with stakeholders in the event of a crisis or contingency. (Enabler: Culture, Ethics and Behaviour) • Includes practices to provide independent assurance over IT risk management. (MEA02)

Figure 45—ISO 31000 Risk Management Framework Covered by *COBIT 5 for Risk (cont.)*	
ISO 31000 Framework Components	***COBIT 5 for Risk* Coverage**
C. Implementing Risk Management	
1. Implementing the framework for managing risk	• Appendix B describes how to implement the seven enablers.
2. Implementing the risk management process	• Includes management practices to develop integrated risk management methods based on an integrated risk management strategy. • Can help enterprises develop leading practice IT risk management techniques in line with the enablers.
D. Monitoring and Review of the Framework	• Contains metrics for achievement of goals and metrics for application of practices.
E. Continual Improvement of the Framework	• Includes management practices and information flows supporting process improvements based on data from incident/event post mortems, adherence to policy and standards, and data on risk-aware culture change.

Process for Managing Risk

ISO 31000 describes six major components of risk management processes. **Figure 46** lists each of these components (labelled A through F) and describes how and where *COBIT 5 for Risk* covers these components.

Figure 46—ISO 31000 Risk Management Processes Covered by *COBIT 5 for Risk*	
ISO 31000 Risk Management Process	***COBIT 5 for Risk* Coverage**
A. Communication and Consultation	• The Information enabler includes specific information to be communicated between the stakeholders.
B. Establishing the Context	
1. Establishing the external context	• Includes management practices to work with the broader enterprise-level risk functions to understand the external context.
2. Establishing the internal context	• Includes management practices to understand the internal context, which include determining where/how organisational process rely on IT for success and comparing to existing IT related capabilities.
3. Establishing the context of the risk management process.	• Features a domain called risk governance to help ensure that the risk management approach adopted is appropriate to the situation of the enterprise and to the risk affecting the achievement of its objectives.
4. Developing risk criteria	• Provides guidance for enterprises to develop their specific risk criteria such as measurement of consequences, defining business impact, establishing risk appetite, and tolerance thresholds and risk aggregation • The enabler model includes management practices to establish risk criteria.
C. Risk Assessment	
1. Risk identification	• Includes management practices to identify the risk associated with key organisational services and products that rely on IT and to identify risk factors that contributed to historical incidents and events. • Includes specific techniques to identify scenarios based on actor, threat type, event, asset/resource and a time dimension.
2. Risk analysis	• Risk analysis is the process whereby frequency and impact of IT risk scenarios are estimated.
3. Risk evaluation	• Addresses this process step intrinsically.
D. Risk Treatment	
1. Selection of options	• Includes guidance on the common response options and how they apply to an IT context.
2. Preparing and implementing risk treatment plans	• Defines specific risk responses to address different treatments for risk (section 2B). • Uses scenario development for risk identification.
E. Monitoring and Review	• Includes goals and metrics that can be used to measure performance and a maturity model to set up a road map for improving risk management processes.
F. Recording the Risk Management Process	• Includes management practices to track key risk decisions and specifies inputs and outputs among its management practices.

Conclusion

COBIT 5 for Risk addresses all ISO 31000 principles, through the *COBIT 5 for Risk* principles and enablers themselves, its conceptual design or through the enabler models. In addition, the framework and process model aspects are covered in greater detail by the *COBIT 5 for Risk* process model. All elements are included in *COBIT 5 for Risk* and are often expanded on or elaborated in greater detail, specifically for IT risk governance and management.

Page intentionally left blank

CHAPTER 2
COBIT 5 FOR RISK AND ISO/IEC 27005

2.1 ISO/IEC 27005:2011—Information Technology—Security Techniques—Information Security Risk Management

ISO/IEC 27005:2011—Information technology— Security techniques—Information security risk management (referred to hereafter as ISO/IEC 27005) defines an information security risk management process that includes the process steps shown in **figure 47**.

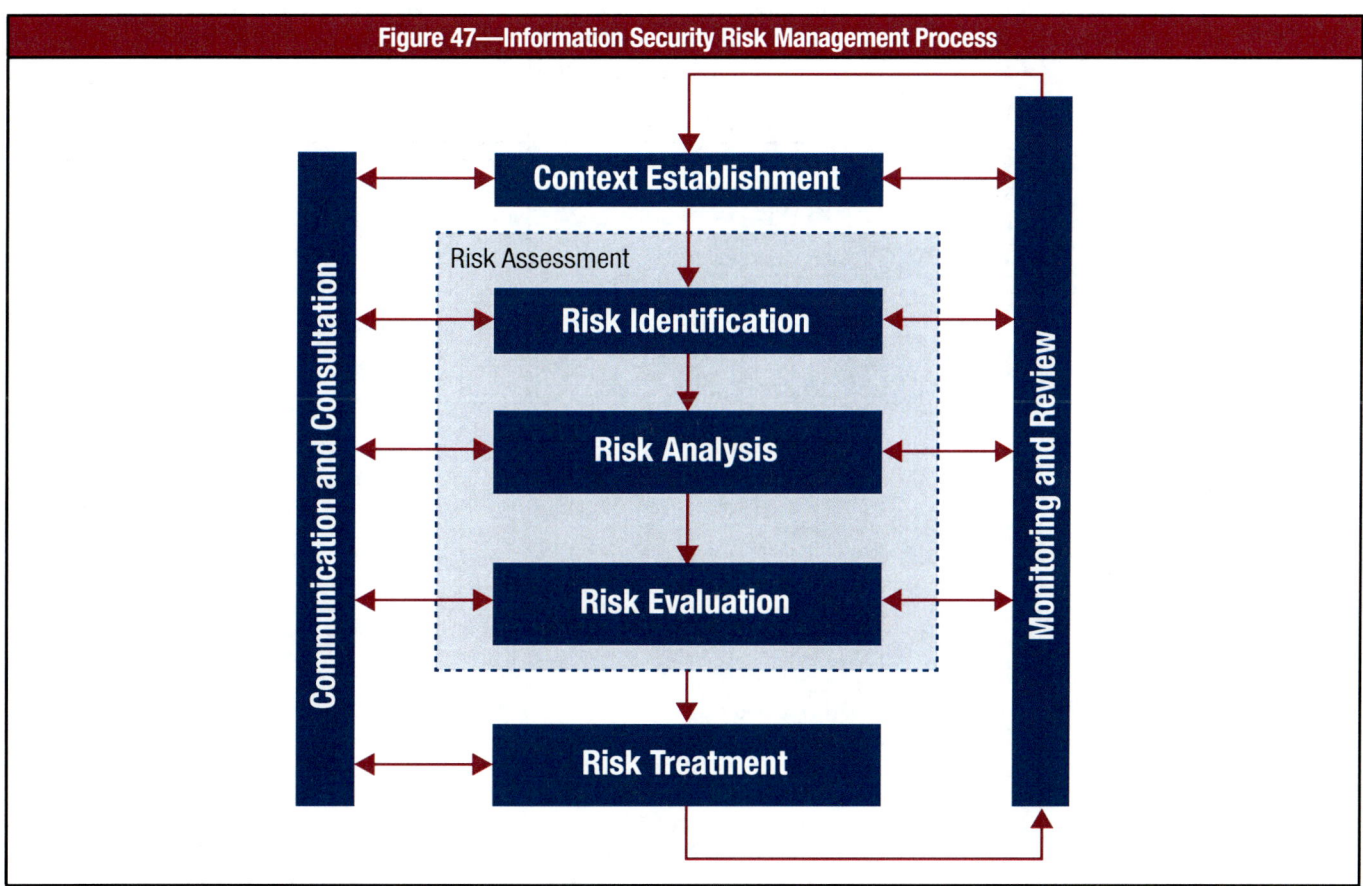

Figure 47—Information Security Risk Management Process

Comparison of ISO/IEC 27005 and COBIT 5 for Risk
Figure 48 highlights the different process steps of ISO/IEC 27005, a summary of the important concepts of these process steps, as well as how and to which extent *COBIT 5 for Risk* covers them.

Overall, it can be noted that the process as defined in ISO/IEC 27005 is fully covered by the different processes and practices of the *COBIT 5 for Risk* process model. The *COBIT 5 for Risk* model provides more extensive guidance and includes areas not covered by ISO/IEC 27005 such as risk governance and reacting to events.

The fundamental difference between the two frameworks is that *COBIT 5 for Risk* addresses a comprehensive number of categories of IT risk, whereas ISO/IEC 27005 focuses specifically on information security risk. ISO/IEC 27005 defines information security risk as 'the potential that threats will exploit vulnerabilities of an information asset or group of information assets and thereby cause harm to an enterprise', whereas the broad definition of IT risk within *COBIT 5 for Risk* is that it is the business risk associated with the use, ownership, operation, involvement, influence and adoption of IT within an enterprise.

Figure 48—ISO/IEC 27005 Process Steps Covered by *COBIT 5 for Risk*		
ISO/IEC 27005 Process Step	**Important Concepts of the Component**	**COBIT 5 for Risk Coverage**
Context Establishment	This process step includes: • Setting the basic criteria necessary for information security risk management (ISRM) • Defining the scope and boundaries • Establishing an appropriate organisation operating the ISRM	This process step is included in the enabler dimension goal. More specifically, contextual quality: the extent to which outcomes of the enabler are fit for purpose given the context in which they operate.
Risk Assessment	Risk assessment determines the value of the information assets, identifies the applicable threats and vulnerabilities that exist (or could exist), identifies the existing controls and their effect on the risk identified, determines the potential consequences, and finally, prioritises the derived risk and ranks it against the risk evaluation criteria set in the context establishment. This process step consists of the following activities: risk identification, risk analysis and risk evaluation.	Appendix B.3 describes risk assessment as one item of the Information enabler.
Risk Identification	Risk identification includes the identification of: • Assets • Threats • Existing Controls • Vulnerabilities • Consequences The output of this process is a list of incident scenarios with their consequences related to assets and business processes.	The sequence used in ISO/IEC 27005 to identify risk is partly aligned to the *COBIT 5 for Risk* approach. The identification of risk comprises the following elements in *COBIT 5 for Risk*: • Control • Value • Threat condition that impose a noteworthy level of IT risk *COBIT 5 for Risk* also uses scenario development for identifying risk. Key attributes of potential and known risk events are stored in a repository. Attributes may include name, description, owner, expected/actual frequency, potential/actual magnitude, potential/actual business impact, disposition, etc.
Risk Analysis	The risk estimation process step includes the following important concepts: • Assessment of consequences • Assessment of incident likelihoods • Determination of level of risk	Risk analysis is the process whereby frequency and impact of IT risk scenarios are estimated.
Risk Evaluation	In this step, levels of risk are compared according to risk evaluation criteria and risk acceptance criteria. The output is a prioritised list of risk elements and the incident scenarios that lead to the identified risk elements.	• Addresses this process step intrinsically as a part of 'risk aggregation'. It evaluates the risk according to 'risk tolerance of the management with regard to the risk appetite of the board'. • Uses a risk map for ranking and graphically displaying risk by defined ranges for frequency and impact.
Risk Treatment	Risk treatment options include: • Risk modification • Risk retention • Risk avoidance • Risk sharing	The treatments of identified risk is described in section 2B. These are: • Risk avoidance • Risk reduction/mitigation • Risk sharing/transfer • Risk acceptance
Information Security Risk Acceptance	The input is a risk treatment plan and the residual risk assessment subject to the risk acceptance criteria. This step comprises the formal acceptance and recording of the suggested risk treatment plans and residual risk assessment by management, with justfication for those that do not meet the enterprise's criteria.	Covers risk acceptance process step in section 2B, subsection 5.2 Risk Response. If an enterprise adopts a risk acceptance stance, it should carefully consider who can accept the risk—even more so with IT risk. IT risk should be accepted only by business management (and business process owners), in collaboration with and supported by IT, and acceptance should be communicated to senior management and the board.

Figure 48—ISO/IEC 27005 Process Steps Covered by *COBIT 5 for Risk* (cont.)		
ISO/IEC 27005 Process Step	**Important Concepts of the Component**	***COBIT 5 for Risk* Coverage**
Information Security Risk Communication and Consultation	This is a transversal process—information about risk should be exchanged and shared between the decision maker and other stakeholders throughout all the steps of the risk management process.	• Principle 1: Meeting stakeholder needs. • The Information enabler includes specific information to be communicated between the stakeholders.
Information Security Risk Monitoring and Review	Risk and its influencing factors should be monitored and reviewed to identify any changes in the context of the organisation at an early stage and to maintain an overview of the complete risk picture.	Includes goals and metrics that can be used to measure performance and a maturity model to set a road map for improving risk management processes.

Conclusion

COBIT 5 for Risk addresses all of the components described within ISO/IEC 27005. Some of the elements are structured or named differently. *COBIT 5 for Risk* takes a broader view on IT risk management compared with ISO/IEC 27005, which is focused on the management of security-related risk. Therefore, there is a strong emphasis in *COBIT 5 for Risk* on processes and practices in order to ensure the alignment with business objectives; the acceptance throughout the enterprise; and the completeness of the scope, amongst other factors.

Page intentionally left blank

CHAPTER 3
COBIT 5 FOR RISK AND COSO ERM

3.1 COSO ERM—Integrated Framework

The Enterprise Risk Management—Integrated Framework of The Committee of Sponsoring Organisations of the Treadway Commission (COSO), also called COSO ERM, defines eight components with regard to the management of business risk. These interrelated components are derived from the way management runs an enterprise and are integrated with the management process. The components are:
• Internal environment
• Objective setting
• Event identification
• Risk assessment
• Risk response
• Control activities
• Information and communication
• Monitoring

Components of COSO ERM
Figure 49 contains the eight components COSO ERM has defined, a summary of the important concepts related to these components, as well as how and to which extent *COBIT 5 for Risk* covers them.

Figure 49—COSO ERM Components Covered by *COBIT 5 for Risk*		
COSO ERM Component	**Important Concepts of the Component**	***COBIT 5 for Risk* Coverage**
Internal environment	This component encompasses the tone of an organization, influencing the risk consciousness of its people, and is the basis for all other components of enterprise risk management, providing discipline and structure. Internal environment factors include an entity's risk management philosophy; its risk appetite; oversight by the board; the integrity, ethical values, and competence of the entity's people; and the way management assigns authority and responsibility, and organizes and develops its people. Hence, this component of ERM is focused on providing guidance to practitioners with regard to risk management and ensuring that ERM is a way of thinking fully embedded in the organization.	The concepts described in the chapter about Internal Environment are inherent throughout the framework: • Principle 2: Covering the enterprise end-to-end. • Principle 4: Enabling a holistic approach defines a set of enablers to support the implementation of a comprehensive governance and management system for enterprise IT. • An elaborated distribution of roles and responsibilities is described in section 2A, chapter 8, Enabler: People, Skills and Competencies, and chapter 4, Enabler: Organisational Structure, of the framework. • Concepts such as risk universe, risk appetite and risk tolerance are described in section 2A, chapter 6, Enabler: Information. • Awareness and communication of risk linking into the creation of a risk-aware culture is described in section 2A, chapter 5, Enabler: Culture, Ethics and Behaviour. • Entity risk management philosophy is described in section 2A, chapter 2, Enabler: Principles, Policies and Frameworks.

Figure 49—COSO ERM Components Covered by *COBIT 5 for Risk (cont.)*		
COSO ERM Component	**Important Concepts of the Component**	***COBIT 5 for Risk* Coverage**
Objective setting	COSO ERM states that objectives are set at the strategic level, establishing a basis for operations, reporting, and compliance objectives. Every entity faces a variety of risk from external and internal sources, and a precondition to effective event identification, risk assessment, and risk response is the establishment of objectives. Objectives are aligned with the entity's risk appetite, which drives risk tolerance levels for the entity.	This component is closely related to risk governance. The following parts of this guide are relevant to objective setting: • Principle 1: Meeting stakeholder needs helps to ensure that enterprise objectives are achieved throughout the goals cascade. • Two out of seven risk principles explained in appendix B.1.1, Risk Principles, are related to setting the objectives: 'Focus on business objectives' and 'Promote fair and open communication'. • Principle 2: Covering the enterprise end-to-end; focuses on integrating governance of enterprise IT (GEIT) into enterprise governance. • Section 2A, chapter 5 covers the awareness and communication of risk. **Figure 63** in appendix B. Risk Communication Plan, gives stakeholder communication information. • Risk appetite and tolerance are discussed in section 2A, chapter 6, Enabler: Information. It is important, however, to acknowledge that within *COBIT 5 for Risk*, enterprise objectives are treated as external input. Enterprise objectives are set at the corporate governance level and stem from stakeholder needs.
Event identification	This component deals with management identifying potential events that, if they occur, will affect the entity, and determines whether they represent opportunities or whether they might adversely affect the entity's ability to successfully implement strategy and achieve objectives. Events with negative impact represent risk, which require management's assessment and response. Events with positive impact represent opportunities, which management channels back into the strategy and the objective-setting process. When identifying events, management considers a variety of internal and external factors that may give rise to risk and opportunities, in the context of the full scope of the organization.	Event identification has been further developed and extended. Events are discussed in the chapters that describe risk scenarios. Developing risk scenarios is a technique of event identification. This guide provides structures, components and guidance on building risk scenarios. The following parts discuss the event identification' component: • Section 2B, The Risk Management Perspective and Using COBIT 5 Enablers, covers the basics of risk scenario development, including event identification. • Appendix D. Using COBIT 5 Enablers to Mitigate IT Risk Scenarios provides guidance on developing scenarios based on specific actors, threat types, events, and assets/resources and a time dimension.
Risk assessment	COSO ERM defines risk assessment as allowing an entity to consider the extent to which potential events have an impact on achievement of objectives. Management assesses events from two perspectives – likelihood and impact– and normally uses a combination of qualitative and quantitative methods. The positive and negative impacts of potential events should be categorised across the entity. Risk is assessed on both inherent and residual basis.	Defines risk assessment as the determination of quantitative or qualitative exposure related to an event. Appendix B.5 describes how each enabler contributes to risk assessment.
Risk response	Having assessed relevant risk, management determines how it will respond. Responses include risk avoidance, reduction, sharing and acceptance. In considering its response, management assesses the effect on risk likelihood and impact, as well as cost and benefits, selecting a response that brings residual risk within desired risk tolerances. Management identifies any opportunities that might be available, and takes an entity-wide, or portfolio, view of risk, determining whether overall residual risk is within the entity's risk appetite.	Section 2B The Risk Management Perspective and Using COBIT 5 Enablers is dedicated to risk response, providing fully developed processes for these activities. It is fully aligned with this COSO ERM component. This process domain has a goal of ensuring that risk issues, opportunities and events are addressed in a cost-effective manner and consistent with business priorities. Appendix B. Detailed Risk Governance and Management Enablers provides hands-on guidance to risk response and prioritisation.

Figure 49—COSO ERM Components Covered by *COBIT 5 for Risk (cont.)*		
COSO ERM Component	**Important Concepts of the Component**	***COBIT 5 for Risk* Coverage**
Control activities	Control activities are the policies and procedures that help ensure that management's risk responses are carried out. Control activities occur throughout the organization, at all levels and in all functions. They include a range of activities as diverse as approvals, authorisations, verifications, reconciliations, reviews of operating performance, security of assets, and segregation of duties. Important concepts in this context are type of activities, policies and procedures, and controls over information systems.	Appendix B details how the enablers can be used to form risk and control activity matrices. Appendix D. Using COBIT 5 Enablers to Mitigate IT Risk Scenarios, links the controls and management practices specified in COBIT 5 to a collection of IT risk scenarios.
Information and communication	Pertinent information is identified, captured and communicated in a form and time frame that enables people to carry out their responsibilities. Information systems use internally generated data and information from external sources, providing information for managing risk and making informed decisions relative to objectives. Effective communication also occurs, flowing down, across and up the organization. There is also effective communication with external parties such as customers, suppliers, regulators and shareholders.	Principle 1: Meeting stakeholder needs provides guidance on awareness and communication, including **figure 2** with different stakeholders, both internal and external to the enterprise. Section 2A, chapter 4, Enabler: Organisational Structures, discusses risk responsibility and accountability. Section 2A, chapter 5, Enabler: Culture, Ethics and Behaviour, discusses promotion of an IT risk-aware culture and encourages effective communication of IT risk.
Monitoring	In this section, COSO ERM deals with risk management being monitored—assessing the presence and functioning of its components over time. This is accomplished through ongoing monitoring activities, separate evaluations or a combination of the two. Ongoing monitoring occurs in the normal course of management activities. The scope and frequency of separate evaluations will depend primarily on an assessment of risk and the effectiveness of ongoing monitoring procedures.	Includes goals and metrics that can be used to measure performance and includes a maturity model to set a road map for improving risk management processes.

Conclusion

COBIT 5 for Risk addresses all of the components defined in COSO ERM, and, for some components, extends the coverage of COSO ERM to the specifics of IT use in the enterprise. Although *COBIT 5 for Risk* focuses less on control, it provides linkages to management practices in the COBIT 5 framework. The essentials for both control and general risk management, as defined in COSO ERM, are present in *COBIT 5 for Risk*, either through the principles themselves, the framework conceptual design, the process model or the additional guidance provided in the framework.

Page intentionally left blank

CHAPTER 4
COMPARISON WITH RISK MARKET REFERENCE SOURCE

4.1 Vocabulary Comparisons: *COBIT 5 for Risk* vs. ISO Guide 73 and COSO ERM

COBIT 5 for Risk and ISO Guide 73 on 'Risk Management—Vocabulary'

ISO/IEC 27005 and ISO 31000 use the ISO Guide 73 'Risk Management—Vocabulary' (the glossary publication overarching the risk management ISO/IEC publications) with regard to defining important concepts. A comparison of the ISO Guide 73 and *COBIT 5 for Risk* definitions is provided in **figure 50**, which is comprised of:
• Column 1: ISO Guide 73 Term
• Column 2: ISO Guide 73 Definition
• Column 3: Disposition of the *COBIT 5 for Risk* definition of the same term (marked as Identical, Implicit, Absent, Equivalent)
• Column 4: The *COBIT 5 for Risk* definition of the term
• Column 5: Comments if relevant or required

Figure 50—Comparison of ISO Guide 73 With *COBIT 5 for Risk* Definitions				
ISO Guide 73 Term	ISO Guide 73 Definition	Disposition in *COBIT 5 for Risk*	Definition in *COBIT 5 for Risk*	Comment
Absolute risk	Level of risk without taking into account existing risk controls	Absent	N/A	This notion corresponds to the concept 'inherent risk'. In general, *COBIT 5 for Risk* does not use this concept.
Consequence	Outcome of an event affecting objectives	Implicit	N/A	*COBIT 5 for Risk* uses the concept 'business impact'.
Event	Occurrence or change of a particular set of circumstances	Equivalent	N/A	The term 'scenario' is used to describe 'things happening', and 'event' is one component of a scenario.
Exposure	Susceptibility to gain or loss, usually quantified in terms of potential impact	Equivalent	N/A	COBIT 5 for Risk uses the terms 'exposure' and 'business impact'.
External context	External environment in which the organization seeks to achieve its objectives	Identical	N/A	As part of the 'risk factors' in the 'risk profile' Information item.
Frequency	Number of occurrences of an event or outcome per defined period of time	Identical	N/A	N/A
Heat map	Overview of the organization's main risks plotted in its risk matrix	Equivalent	N/A	*COBIT 5 for Risk* uses the term 'risk map'.
Incident	Event in which a loss occurred or could have occurred regardless of severity	Implicit	N/A	The term 'loss' is used instead.

ISO Guide 73 Term	ISO Guide 73 Definition	Disposition in COBIT 5 for Risk	Definition in COBIT 5 for Risk	Comment
Figure 50—Comparison of ISO Guide 73 With COBIT 5 for Risk Definitions (cont.)				
Internal context	Internal environment in which the organization seeks to achieve its objectives	Equivalent	N/A	Internal context is included as part of the 'risk factors' in the 'risk profile' Information item.
Level of risk	Magnitude of a risk expressed in terms of the combination of consequences and their likelihood	Implicit	N/A	The magnitude of risk is discussed in the framework, where risk appetite, risk reaction, etc., are discussed.
Likelihood	Chance of something happening	Absent	N/A	COBIT 5 for Risk uses the term 'frequency', which allows more accurate assessment of events occurring more than once in a given period.
Probability	Measure of the chance of occurrence expressed as a number between 0 and 1, where 0 is impossibility and 1 is absolute certainty	Equivalent	N/A	Instead, the term 'frequency' is used.
Residual risk	Risk remaining after risk treatment	Equivalent	N/A	The framework does not use 'inherent' or 'absolute risk'. Instead it uses the term 'current risk' and 'residual risk', both taking into account the current and planned controls.
Risk	Effect of uncertainty on objectives	Equivalent	The potential that a given threat will exploit vulnerabilities of an asset or group of assets to cause loss of/or damage to the assets. Or the potential of business objectives not being met.	Definitions are different but equivalent—both contain the concept 'uncertainty'.
Risk aggregation	Process to identify and illustrate the interaction of several, differently correlated individual risk elements of an organization in order to obtain the overall risk	Identical	N/A	N/A
Risk analysis	Process to comprehend the nature of risk and to determine the level of risk	Identical	N/A	N/A
Risk assessment	Overall process of risk identification, risk analysis and risk evaluation	Identical	N/A	N/A
Risk avoidance	Decision not to become involved in, or action to withdraw from, a risk situation	Identical	N/A	N/A
Risk communication	Continual and iterative processes that an organization conducts to provide, share or obtain information, and to engage in dialogue with stakeholders regarding the management of risk	Implicit	N/A	Communication on risk is an important part of COBIT 5 for Risk.

ISO Guide 73 Term	ISO Guide 73 Definition	Disposition in COBIT 5 for Risk	Definition in COBIT 5 for Risk	Comment
Figure 50—Comparison of ISO Guide 73 With *COBIT 5 for Risk* Definitions *(cont.)*				
Risk control	Measure that is modifying risk	Equivalent	N/A	*COBIT 5 for Risk* primarily links the IT risk of an enterprise to COBIT 5 enablers and related goals, metrics and (detailed) activities (see section 2B). COBIT 5 prefers not to use the word 'control' but rather 'enabler' and 'management and governance practices'.
Risk criteria	Terms of reference against which the significance of a risk is evaluated	Equivalent	N/A	*COBIT 5 for Risk* core risk processes include practices and activities to develop and maintain risk criteria.
Risk evaluation	Process of comparing the results of risk analysis with risk criteria to determine whether the risk and/or its magnitude is acceptable or tolerable	Equivalent	N/A	An equivalent process exists in *COBIT 5 for Risk* to describe the estimation of impact and frequency as mentioned previously.
Risk identification	Process of finding, recognizing and describing risk.	Implicit	N/A	As mentioned in the previous comparison, *COBIT 5 for Risk* uses the scenario technique as a practical approach for risk identification.
Risk management	Coordinated activities to direct and control an organization with regard to risk	Implicit	N/A	The term 'risk management' is used holistically in order to cover all concepts and (governance and management) processes affiliated with managing risk.
Risk management process	Systematic application of management policies, procedures and practices to the tasks of communicating, consulting, establishing the context, identifying, analyzing, evaluating, treating, monitoring and reviewing risk	Implicit	N/A	See 'risk management'.
Risk matrix	Tool for ranking and displaying risk by defining risk categories (e.g., financial risk, safety risk, contextual risk, etc.) and defining ranges for consequences and levels of likelihood for each category	Identical	N/A	*COBIT 5 for Risk* uses the term 'risk map'. See also 'heat map'.
Risk owner	Person with the authority and accountability to make a decision to treat, or not to treat a risk	Equivalent	N/A	RACI charts in the *COBIT 5 for Risk* Organisational Structures enabler assign risk owners.
Risk profile	Description of an organization's risk	Identical	N/A	N/A
Risk register	Record of information about identified risk	Identical	N/A	N/A
Risk sharing	Form of risk treatment involving the agreed distribution of risk with other parties	Identical	N/A	*COBIT 5 for Risk* uses 'risk transfer' in combination with 'risk sharing'. The sharing of risk is a consequence of transferring risk to other parties.
Risk tolerance	Organization's readiness to accept a residual risk after risk treatment in order to achieve the organization's objectives	Equivalent	The acceptable level of variation that management is willing to allow for any particular risk as it pursues objectives	*COBIT 5 for Risk* distinguishes between 'risk appetite' and 'risk tolerance'. (See the glossary.)

Figure 50—Comparison of ISO Guide 73 With *COBIT 5 for Risk* Definitions *(cont.)*				
ISO Guide 73 Term	**ISO Guide 73 Definition**	**Disposition in *COBIT 5 for Risk***	**Definition in *COBIT 5 for Risk***	**Comment**
Risk treatment	Process to modify risk	Identical	N/A	*COBIT 5 for Risk* uses 'risk reduction' in combination with 'risk mitigation'. Both terms can be used interchangeably.
Uncertainty	State, even partial, of deficiency of information related to, understanding or knowledge of, an event, its consequence or likelihood	Absent	N/A	The term as such is absent. Rather, *COBIT 5 for Risk* describes the different risk factors enabling or creating uncertainty.
Vulnerability	A weakness of an asset or group of assets that can be exploited by one or more threats	Equivalent	N/A	The term 'vulnerability' is used in *COBIT 5 for Risk*. Also, when analysing risk scenarios, COBIT 5 considers 'risk factors', which conveys the same meaning.

COBIT 5 for Risk and COSO ERM on 'Risk Management—Vocabulary'

Both COSO ERM and *COBIT 5 for Risk* have defined a number of terms. The exact definition or sometimes meaning of these terms can differ between both frameworks. To understand these differences, see **figure 51**, which lists:
• Column 1: COSO ERM Term
• Column 2: COSO ERM Definition
• Column 3: Disposition of the *COBIT 5 for Risk* definition of the same term (marked as Identical, Implicit, Absent, Equivalent)
• Column 4: The *COBIT 5 for Risk* definition of the term
• Column 5: Comments if relevant or required

The purpose is to provide information about COSO ERM with a comparison to *COBIT 5 for Risk* starting from a known basis. In addition, this figure helps to prevent purely semantic discussions. Due to the control-focused character of COSO ERM, only the relevant risk-related concepts are compared. The control related concepts are linked to the other ISACA frameworks and are therefore not in this list.

Figure 51—Comparison of COSO ERM and *COBIT 5 for Risk* Definitions				
COSO ERM Term	**COSO ERM Definition**	**Disposition in *COBIT 5 for Risk***	**Definition in *COBIT 5 for Risk***	**Comment**
Criteria	A set of standards against which enterprise risk management can be measured in determining effectiveness. The eight enterprise risk management components, taken in the context of inherent limitations of enterprise risk management, represent criteria for enterprise risk management effectiveness for each of the four objectives categories.	Implicit	N/A	A more practical approach for measuring risk management effectiveness can be found in the metrics and capability model of COBIT 5.
Deficiency	A condition within enterprise risk management worthy of attention that may represent a perceived, potential, or real shortcoming, or an opportunity to strengthen enterprise risk management to provide a greater likelihood that the entity's objectives will be achieved	Implicit	N/A	*COBIT 5 for Risk* more often uses the term 'risk factor'.
Design	1. Intent. As used in the definition, enterprise risk management is intended to identify potential events that may affect the entity, and manage risk to be within its risk appetite, to provide reasonable assurance as to achievement of objectives. 2. Plan. The way a process is supposed to work, contrasted with how it actually works.	Implicit	N/A	N/A

	Figure 51—Comparison of COSO ERM and *COBIT 5 for Risk* Definitions *(cont.)*				
COSO ERM Term	**COSO ERM Definition**	**Disposition in *COBIT 5 for Risk***	**Definition in *COBIT 5 for Risk***	**Comment**	
Effected	Used with enterprise risk management: devised and maintained	Absent	N/A	N/A	
Enterprise risk management process	A synonym for enterprise risk management applied in an entity	Implicit	N/A	The term 'risk management' is used holistically in order to cover all concepts and processes affiliated with managing risk.	
Event	An incident or occurrence, from sources internal or external to an entity, which affects achievement of objectives	Implicit	IT risk scenario: The description of an IT-related event that can lead to a business impact.	The term scenario is used to describe 'things happening'.	
Impact	Result or effect of an event. There may be a range of possible impacts associated with an event. The impact of an event can be positive or negative relative to the entity's related objectives.	Identical	N/A	N/A	
Inherent limitations	Those limitations of enterprise risk management. The limitations relate to the limits of human judgment; resource constraints, and the need to consider the cost of controls in relation to expected benefits; the reality that breakdowns can occur; and the possibility of management override and collusion.	Implicit	N/A	N/A	
Inherent risk	The risk to an entity in the absence of any actions management might take to alter either the risk's likelihood or impact	Absent	N/A	*COBIT 5 for Risk* does not make use of 'inherent' or 'absolute risk'. It rather works with 'residual risk', calling it 'risk'.	
Likelihood	The possibility that a given event will occur. Terms sometimes take on more specific connotations, with 'likelihood' indicating the possibility that a given event will occur in qualitative terms such as high, medium, and low, or other judgmental scales, and 'probability' indicating a quantitative measure such as a percentage, frequency of occurrence, or other numerical metric.	Absent	N/A	*COBIT 5 for Risk* uses the term 'frequency', which allows for a more accurate assessment of events occurring more than once in a given period.	
Management intervention	Management's actions to overrule prescribed policies or procedures for legitimate purposes; management intervention is usually necessary to deal with non-recurring and non-standard transactions or events that otherwise might be handled inappropriately by the system (contrast this term with 'management override')	Absent/Implicit	N/A	Not part of the *COBIT 5 for Risk* framework per se. The COBIT 5 framework itself provides for a governance and management framework of which principles, policies, responsibilities and business controls are all part.	
Management override	Management's overruling of prescribed policies or procedures for illegitimate purposes with the intent of personal gain or an improperly enhanced presentation of an entity's financial condition or compliance status (contrast this term with Management Intervention).	Absent	N/A	Not part of the *COBIT 5 for Risk* framework per se. The COBIT 5 framework itself provides for a governance and management framework of which principles, policies, responsibilities and business controls are all part.	
Management process	The series of actions taken by management to run an entity. Enterprise risk management is a part of an integrated management process	Implicit	N/A	*COBIT 5 for Risk* defines a process outlining the approach to integrate with the ERM management process.	

	Figure 51—Comparison of COSO ERM and *COBIT 5 for Risk* Definitions *(cont.)*				
COSO ERM Term	**COSO ERM Definition**	**Disposition in** *COBIT 5 for Risk*	**Definition in** *COBIT 5 for Risk*	**Comment**	
Opportunity	The possibility that an event will occur and positively affect the achievement of objectives	Identical	N/A	The upside of risk is also acknowledged in *COBIT 5 for Risk*.	
Policy	Management's dictate of what should be done to effect control. A policy serves as the basis for procedures for its implementation.	Identical	N/A	The term 'policies' as described in the enabler Principles, Policies and Frameworks maps to this term.	
Procedure	An action that implements a policy	Identical	N/A	The term 'procedure' as described in the enabler Principles, Policies and Frameworks maps to this term.	
Reporting	Used with 'objectives': having to do with the reliability of the entity's reporting, including both internal and external reporting of financial and non-financial information	Implicit	N/A	*COBIT 5 for Risk* focuses extensively on information items, used for reporting, in the enabler Information.	
Residual risk	The remaining risk after management has taken action to alter the risk's likelihood or impact	Absent	N/A	*COBIT 5 for Risk* does not make use of 'inherent' or 'absolute risk'. It rather works with 'residual risk', calling them 'risk'.	
Risk	The possibility that an event will occur and adversely affect the achievement of objectives	Equivalent	1. The potential that a given threat will exploit vulnerabilities of an asset or group of assets to cause loss of/or damage to the assets 2. The potential of business objectives not being met 3. The combination of the probability of an event and its consequence	Definitions are different but equivalent. COBIT 5 also talks about the positive side of risk.	
Risk appetite	The broad-based amount of risk an organisation is willing to accept in pursuit of its mission (or vision)	Equivalent	The amount of risk, on a broad level, an entity is willing to accept in pursuit of its mission. Risk appetite is defined in the context of a mission statement or strategy formulation and will therefore be more qualitative than risk tolerance.	N/A	
Risk tolerance	The acceptable variation relative to the achievement of an objective	Equivalent	The acceptable level of variation that management is willing to allow for any particular risk as the enterprise pursues its objectives	N/A	
Stakeholders	Parties that are affected by the entity, such as shareholders, the communities in which the entity operates, employees, customers and suppliers	Implicit	N/A	Principle 1: Meeting Stakeholders Needs and enabler Organisational Structures cover the term 'stakeholders' in the COBIT 5 for Risk publication.	
Uncertainty	Inability to know in advance the exact likelihood or impact of future events	Absent	N/A	The term as such is absent. The publication rather describes the different factors enabling or creating uncertainty.	

APPENDIX A
GLOSSARY

Term	Explanation
Asset	Something of either tangible or intangible value that is worth protecting, including people, systems, infrastructure, finances and reputation
Business goal	The translation of the enterprise's mission from a statement of intention into performance targets and results
Business impact	The net effect, positive or negative, on the achievement of business objectives
Business impact analysis (BIA)	Evaluating the criticality and sensitivity of information assets. An exercise that determines the impact of losing the support of any resource to an enterprise, establishes the escalation of that loss over time, identifies the minimum resources needed to recover, and prioritises the recovery of processes and the supporting system.
Business objective	A further development of the business goals into tactical targets and desired results and outcomes
Enterprise risk management (ERM)	The discipline by which an enterprise in any industry assesses, controls, exploits, finances, and monitors risk from all sources for the purpose of increasing the enterprise's short- and long-term value to its stakeholders
Event	Something that happens at a specific place and/or time
Event type	For the purpose of IT risk management,[7] one of three possible sorts of events: threat event, loss event and vulnerability event
Frequency	A measure of the rate by which events occur over a certain period of time
IT risk	The business risk associated with the use, ownership, operation, involvement, influence and adoption of IT within an enterprise
IT risk profile	A description of the overall (identified) IT risk to which the enterprise is exposed
IT risk register	A repository of the key attributes of potential and known IT risk issues. Attributes may include name, description, owner, expected/actual frequency, potential/actual magnitude, potential/actual business impact and disposition.
IT risk scenario	The description of an IT-related event that can lead to a business impact
IT-related incident	An IT-related event that causes an operational, developmental and/or strategic business impact
Key risk indicator (KRI)	A subset of risk indicators that are highly relevant and possess a high probability of predicting or indicating important risk
Lag indicator	Metrics for achievement of goals—An indicator relating to the outcome or result of an enabler, i.e., this indicator is only available after the facts or events
Lead indicator	Metrics for application of good practice—An indicator relating to the functioning of an enabler, i.e., this indicator will provide an indication on possible outcome of the enabler
Loss event	Any event during which a threat event results in loss
Magnitude	A measure of the potential severity of loss or the potential gain from realised events/scenarios
Residual risk	The remaining risk after management has implemented a risk response
Risk (business)	A probable situation with uncertain frequency and magnitude of loss (or gain)
Risk aggregation	The process of integrating risk assessments at a corporate level to obtain a complete view on the overall risk for the enterprise
Risk analysis	1. A process by which frequency and magnitude of IT risk scenarios are estimated 2. The initial steps of risk management: analysing the value of assets to the business, identifying threats to those assets and evaluating how vulnerable each asset is to those threats
Risk appetite	The amount of risk, on a broad level, an entity is willing to accept in pursuit of its mission
Risk assessment	A process used to identify and evaluate risk and its potential effects
Risk culture	The set of shared values and beliefs that governs attitudes toward risk-taking, care and integrity, and determines how openly risk and losses are reported and discussed
Risk factor	A condition that can influence the frequency and/or magnitude and, ultimately, the business impact of IT-related events/scenarios
Risk indicator	A metric capable of showing that the enterprise is subject to, or has a high probability of being subject to, a risk that exceeds the defined risk appetite
(IT) Risk issue	1. An instance of an IT risk 2. A combination of control, value and threat conditions that impose a noteworthy level of IT risk
Risk map	A (graphic) tool for ranking and displaying risk by defined ranges for frequency and magnitude

[7] Being able to consistently and effectively differentiate the different types of events that contribute to risk is a critical element in developing good risk-related metrics and well-informed decisions. Unless these categorical differences are recognized and applied, any resulting metrics lose meaning and, as a result, decisions based on those metrics are far more likely to be flawed.

Term	Explanation
Risk response	Risk avoidance, risk acceptance, risk sharing/transfer, risk mitigation, leading to a situation that as much future residual risk (current risk with the risk response defined and implemented) as possible (usually depending on budgets available) falls within risk appetite limits
Risk statement	A description of the current conditions that may lead to the loss; and a description of the loss. Source: Software Engineering Institute (SEI). For a risk to be understandable, it must be expressed clearly. Such a statement must include a description of the current conditions that may lead to the loss; and a description of the loss.
Risk tolerance	The acceptable level of variation that management is willing to allow for any particular risk as the enterprise pursues its objectives
Threat	Anything (e.g., object, substance, human) that is capable of acting against an asset in a manner that can result in harm
Threat event	Any event during which a threat element/actor acts against an asset in a manner that has the potential to directly result in harm
Vulnerability	A weakness in the design, implementation, operation or internal control of a process that could expose the system to adverse threats from threat events
Vulnerability event	Any event during which a material increase in vulnerability results. Note that this increase in vulnerability can result from changes in control conditions or from changes in threat capability/force.

APPENDIX B
DETAILED RISK GOVERNANCE AND MANAGEMENT ENABLERS

Appendix B provides more detailed guidance on the risk governance and management enablers, as introduced in section 2A of this guide. This appendix can be used to gain a deeper understanding of the content and meaning of each of the elements of the enablers.

B.1 Enabler: Principles, Policies and Frameworks

Figure 52 provides risk principles.

	Figure 52—Risk Principles	
Ref.	**Principle**	**Explanation**
1	Connect to enterprise objectives	Enterprise objectives and the amount of risk that the enterprise is prepared to take are clearly defined and drive IT risk management.
2	Align with ERM	IT risk is treated as a business risk, as opposed to a separate type of risk, and the approach is comprehensive and cross-functional.
3	Balance cost/benefit of IT risk	Risk is prioritised and addressed in line with risk appetite and tolerance.
4	Promote fair and open communication	Open, accurate, timely and transparent information on IT risk is exchanged and serves as the basis for all risk-related decisions.
5	Establish tone at the top and accountability	Key people, i.e., influencers, business owners and the board, are engaged in IT risk management and take culture and behaviour into account. They make informed decisions with appropriate accountabilities based on best available information. Explicitly addresses uncertainty.
6	Function as part of daily activities	Risk management practices are appropriately prioritised and embedded in enterprise decision-making processes.
7	Consistent approach	Risk management practices are applied continually and are improved, enhanced and aligned.

B.1.2 Risk Policy
Figure 53 is a potential table of contents for a risk policy. Details are provided for each of the items, often referring to other sections in this publication.

```
Figure 53—Risk Policy Table of Contents Example

Table of Contents
1. Scope
2. Validity
3. Management Commitment and Accountability
4. Risk Governance
        4.1 Principles
        4.2 Evaluate
                4.2.1 Stakeholder Needs
                4.2.2 Drivers and Goals
        4.3 Direct
                4.3.1 Enterprise
                4.3.2 Roles and Responsibilities
                4.3.3 Objectives
        4.4 Monitor
                4.4.1 Metrics
                4.4.2 Communication
5. Risk Management Framework
```

Additional Guidance on Some Chapters

1. SCOPE
The following text provides an example of a scope declaration in a risk policy.

A risk policy comprises governing and managing all externally and internally driven as well as strategic, delivery, and operational risk. It helps to demonstrate a good practice of risk governance and management, earn competitive advantage and support business objectives. Moreover, risk governance and management is also viewed as an essential component of corporate governance, strongly supported by the organisation's internal control system and value charter.

To be effective, risk management must be embedded in the normal process and form part of the daily management practice. This will lead to sound, risk-based decision making, establish a risk-aware culture among all employees at all levels and provide the necessary assurance to stakeholders.

The purpose of this policy is to deploy a risk management process within corporate units in accordance with the IT strategic plan, which provides the vision, objectives and principles of IT to be applied throughout the enterprise and the basic guidelines on how to apply them in practice. The objective of this policy is to build a sustainable IT risk framework that supports managing IT risk in a cost-effective and pragmatic way, while complying with requirements.

2. VALIDITY
Figure 54 describes the three aspects of validity that should be clearly identified in a risk policy.

Figure 54—Validity Aspects to be Identified in a Risk Policy	
Validity Aspect	**Description**
Applicable	States to whom in the enterprise that the policy is applicable
Update and revalidation	States the frequency of updating the policy and the persons from the enterprise who will revalidate the updated policy
Distribution	States the distribution list of the risk policy. A difference can be made between the people to whom the policy is explicitly communicated and to whom the policy is available. It is the responsibility of top management to make sure that the distribution happens appropriately, with a minimum distribution to which the policy is applicable.

3. MANAGEMENT COMMITMENT AND ACCOUNTABILITY
Probably the most important influencing factor in effective risk management is the demonstration of executive managements' support for the risk management programme. This means that reasonable consideration needs to be given and appropriate action should be taken on viable proposals or recommendations.

For that reason, holding management at all levels of the hierarchy accountable for the fair and consistent application of the risk management practices is a key influencing factor. This commitment and accountability should be clearly communicated and detailed in the risk policy.

4. RISK GOVERNANCE
4.1 **Principles**—Risk governance principles are listed in the section on risk principles (B.1.1).
4.2 **Evaluate**: Determine balanced, agreed-on enterprise objectives to be achieved.
    4.2.1 **Stakeholder Needs**—Stakeholders for risk management need to be defined. **Figure 2** identifies all potential stakeholders and their interest. Based on that information, this section in the policy can be completed. Stakeholder needs can be addressed also in terms of COBIT 5 enterprise goals and IT-related goals, as defined in the COBIT 5 framework.
    4.2.2 **Drivers and Goals**—The goals of risk management should be defined based on the needs of stakeholders listed above and on the enterprise goals and IT-related goals.
4.3 **Direct**—Direction of risk management is expressed in activities, accountability, and roles and responsibilities by objectives:
    4.3.1 **Enterprise**—The organisational structure(s) put in place to manage risk should be described; the Organisational Structures (section 2A, chapter 4) can be used for that purpose. Appendix B.3 contains a much more detailed description of these organisational structures, their key practices and how they can be implemented. This section can be used to define this part of the risk policy.
    4.3.2 **Roles and Responsibilities**—Should be established where key responsibilities are defined. Overall accountability should be clearly assigned to the board. Practical guidance in this respect can be found in section 2A, chapter 4 and appendices B.2 and B.3, where RACI charts and responsibilities are discussed. The three lines of defence principle also apply here.
    4.3.3 **Objectives**—The key objectives of risk management should be defined and aligned with the goals in section 2.3. The goals should be SMART (specific, measurable, attainable, realistic and timely) and should address efficiency and effectiveness of risk management and related processes.
4.4 **Monitor**—Risk activity should be monitored based on standard risk reporting that allows decision making.
    4.4.1 **Metrics**—The appropriate metrics should be defined for the measurement of risk management and related processes. COBIT 5 contains a generic enabler performance model, and all metrics can be based on this model. For example:
        • All metrics related to the achievement of enabler goals, e.g., process goals, organisational structure goals
        • All metrics related to the application of good practice for enablers, e.g., application of process practices
        • All metrics related to the life cycle management of enablers
        • Combined metrics, e.g., process capability levels (according to ISO/IEC 15504-based process assessments)
        • KRIs

4.4.2 **Communication**—Risk capabilities, risk status and risk profile should be maintained and communicated to all relevant stakeholders. Various communication mechanisms are identified in this guide, for example:
- •Relevant risk management processes
- • Appropriate information items—the most important here is the risk profile; the risk policy could, for example, identify the key components the risk profile should contain and how often these should be made available.

5. RISK MANAGEMENT FRAMEWORK

The risk management framework defines—at a high level—all enablers the enterprise will put in place for their risk management efforts. In that respect, this description can be based on section 2A in its entirety.

Page intentionally left blank

B.2. Enabler: Processes

This section contains more detailed guidance on the Process enabler for risk governance and management, i.e., more details on those processes that are important to build and sustain an effective and efficient risk function in an enterprise.

For each key supporting process, the following information is provided:
• Process description and process purpose statement
• Risk-specific process goals and metrics
• Risk-specific process practices, inputs and outputs, and process activities

B.2.1 Key Supporting Processes

The processes listed in **figure 55** are key supporting processes for building an effective and efficient risk function in the enterprise, as identified in section 2A, subsection 3.2.

Figure 55—Risk Function Key Supporting Processes		
Process Identification	**Justification**	**Output**
EDM01 Ensure Governance Framework Setting and Maintenance	Governing and managing risk requires the setup of an adequate governance framework, to put in place enabling structures, principles, processes and practices.	Risk governance guiding principles
EDM02 Ensure Benefits Delivery	This process focuses on managing the value that the risk function generates.	Actions to improve risk value delivery
EDM05 Ensure Stakeholder Transparency	The enterprise risk function requires transparent performance and conformance measurement, with goals and metrics approved by stakeholders.	Evaluation of risk reporting requirements
APO02 Manage Strategy	IT risk management strategy must be well defined and aligned to ERM approach.	Risk management strategy
APO06 Manage Budget and Costs	The risk function needs to be budgeted.	Financial and budgetary requirements
APO07 Manage Human Resources	Risk management requires the right amount of people, skills and experience.	HR competencies framework
APO08 Manage Relationships	Maintain the relationships between the risk function and the business.	Communication plan
APO11 Manage Quality	Quality is an essential component of an effective risk management.	Quality review of risk deliverables
BAI08 Manage Knowledge	The risk function needs to be provided with the knowledge required to support staff in their work activities.	Classification of risk function information, access control over information, rules for disposal of information
MEA01 Monitor, Evaluate and Assess Performance and Conformance	Risk is a key aspect in the monitoring, evaluating and assessing of business and IT.	Risk monitoring metrics and targets
MEA02 Monitor, Evaluate and Assess the System of Internal Control	Internal controls are key in monitoring and containing risk, to avoid risk becoming an issue.	Results of internal control monitoring and reviews
MEA03 Monitor, Evaluate and Assess Compliance With External Requirements	Compliance with laws, regulations and contractual requirements represent risk and have to be monitored, evaluated and assessed in alignment with enterprise strategy.	Reports of non-compliance issues and root causes

Page intentionally left blank

EDM01 Ensure Governance Framework Setting and Maintenance	Area: Governance Domain: Evaluate, Direct and Monitor

COBIT 5 Process Description
Analyse and articulate the requirements for the governance of enterprise IT, and put in place and maintain effective enabling structures, principles, processes and practices, with clarity of responsibilities and authority to achieve the enterprise's mission, goals and objectives.

COBIT 5 Process Purpose Statement
Provide a consistent approach integrated and aligned with the enterprise governance approach. To ensure that IT-related decisions are made in line with the enterprise's strategies and objectives, ensure that IT-related processes are overseen effectively and transparently, compliance with legal and regulatory requirements is confirmed, and the governance requirements for board members are met.

EDM01 Risk-specific Process Goals and Metrics

Risk-specific Process Goals	Related Metrics
1. The risk governance system is embedded in the enterprise.	• Number of enterprise and IT processes in which risk activities are integrated • Degree by which agreed-on IT risk governance principles are evidenced in processes and practices (percentage of processes and practices with clear traceability to principles)
2. Assurance is obtained over the risk governance system.	• Frequency of independent reviews of IT risk governance documentation • Number of IT risk governance issues reported • Frequency of IT risk governance reporting to the executive committee and board • Number of external/internal audits and reviews

EDM01 Risk-specific Process Practices, Inputs/Outputs and Activities

Governance Practice	Risk-specific Inputs (in Addition to COBIT 5 Inputs)		Risk-specific Outputs (in Addition to COBIT 5 Outputs)	
	From	Description	Description	To
EDM01.01 Evaluate the governance system. Continually identify and engage with the enterprise's stakeholders, document an understanding of the requirements, and make a judgement on the current and future design of governance of enterprise IT.	Outside *COBIT 5 for Risk*	Internal and external environmental factors (legal, regulatory and contractual obligations) and trends	Decision-making model for IT risk	EDM01.02 EDM01.03 EDM02.03
			Stakeholder requirements with regard to risk priorities and objectives	EDM01.03 EDM02.01 EDM05.02 MEA01.01

Risk-specific Activities (in Addition to COBIT 5 Activities)

1. Analyse and identify the internal and external environmental risk factors (legal, regulatory and contractual obligations) and trends in the business and IT environment that may influence risk governance design.
2. Determine the significance of IT risk and its role with respect to the business.
3. Determine the IT risk implications as a result of the overall enterprise control environment.
4. Identify relevant stakeholders with regard to governance of risk (business process owners, management, CRO, etc.)
5. Collect stakeholder requirements with regard to risk priorities and objectives.
6. Articulate IT risk decision-making in alignment with the enterprise governance design principles.
7. Understand the enterprise's decision-making culture and determine the optimal decision-making model for IT risk.
8. Determine the appropriate levels of authority delegation, including threshold rules, for IT risk decisions.
9. Coach management decision makers on the proposed IT risk analysis approach. Illustrate how risk analysis results can benefit major decisions. Describe what level of quality decision makers should expect, how to interpret risk analysis reports, definitions of key terms, and the limitations of measurements and estimates based on incomplete data. Identify gaps with enterprise risk expectations.

EDM01 Risk-specific Process Practices, Inputs/Outputs and Activities *(cont.)*				
	Risk-specific Inputs (in Addition to COBIT 5 Inputs)		**Risk-specific Outputs (in Addition to COBIT 5 Outputs)**	
Governance Practice	**From**	**Description**	**Description**	**To**
EDM01.02 Direct the governance system. Inform leaders and obtain their support, buy-in and commitment. Guide the structures, processes and practices for the governance of IT in line with agreed-on governance design principles, decision-making models and authority levels. Define the information required for informed decision making.	EDM01.01	Decision-making model for IT risk	• Risk function mandate • IT risk governance principles	Internal

Risk-specific Activities (in Addition to COBIT 5 Activities)
1. Communicate IT risk governance principles and agree with executive management on the way forward to establish informed and committed leadership.   2. Establish or delegate the establishment of IT risk governance structures, processes and practices in line with agreed-on design principles.   3. Allocate responsibility, authority and accountability in line with agreed-on IT risk governance design principles, decision-making models and delegation. Mandate an enterprisewide risk function.   4. Ensure that IT risk communication and reporting mechanisms provide those responsible for oversight and decision-making with appropriate information in a timely manner.

	Risk-specific Inputs (in Addition to COBIT 5 Inputs)		**Risk-specific Outputs (in Addition to COBIT 5 Outputs)**	
Governance Practice	**From**	**Description**	**Description**	**To**
EDM01.03 Monitor the governance system. Monitor the effectiveness and performance of the enterprise's governance of IT. Assess whether the governance system and implemented mechanisms (including structures, principles and processes) are operating effectively and provide appropriate oversight of IT.	EDM01.01	• Decision-making model for IT risk • Stakeholder requirements with regard to risk priorities and objectives	• Assessment of IT risk governance mechanisms • Formal meeting minutes and communication of action plans	Internal

Risk-specific Activities (in Addition to COBIT 5 Activities)
1. Periodically assess whether agreed-on IT risk governance mechanisms (structures, principles, processes, etc.) are established and operating effectively.   2. Identify actions to rectify any deviations found.   3. Monitor IT and business processes to ensure that these comply with IT risk activities.   4. Document risk governance decision-making through formal meeting minutes and communication of action plans.

EDM02 Ensure Benefits Delivery	Area: Governance Domain: Evaluate, Direct and Monitor

COBIT 5 Process Description
Optimise the value contribution to the business from the business processes, IT services and IT assets resulting from investments made by IT at acceptable costs.

COBIT 5 Process Purpose Statement
Secure optimal value from IT-enabled initiatives, services and assets; cost-efficient delivery of solutions and services; and a reliable and accurate picture of costs and likely benefits so that business needs are supported effectively and efficiently.

EDM02 Risk-specific Process Goals and Metrics

Risk-specific Process Goals	Related Metrics
1. Benefits and cost of the risk function are balanced and managed and contribute optimum value.	• Percent of risk reduction vs. budget deviation (budgeted vs. projection) • Level of stakeholder satisfaction with the risk management measures in place, based on surveys

EDM02 Risk-specific Process Practices, Inputs/Outputs and Activities

Governance Practice	Risk-specific Inputs (in Addition to COBIT 5 Inputs)		Risk-specific Outputs (in Addition to COBIT 5 Outputs)	
	From	Description	Description	To
EDM02.01 Evaluate value optimisation. Continually evaluate the portfolio of IT-enabled investments, services and assets to determine the likelihood of achieving enterprise objectives and delivering value at a reasonable cost. Identify and make judgement on any changes in direction that need to be given to management to optimise value creation.	EDM01.01	Stakeholder requirements with regard to risk priorities and objectives	Formal documentation of stakeholder requirements and direction with regard to risk tolerance levels in IT investment policy	EDM02.02 EDM02.03

Risk-specific Activities (in Addition to COBIT 5 Activities)

1. Set direction in accordance with the requirements of stakeholders (such as shareholders, regulators, auditors and customers) for protecting their interests and delivering value.

Governance Practice	Risk-specific Inputs (in Addition to COBIT 5 Inputs)		Risk-specific Outputs (in Addition to COBIT 5 Outputs)	
	From	Description	Description	To
EDM02.02 Direct value optimisation. Direct value management principles and practices to enable optimal value realisation from IT-enabled investments throughout their full economic life cycle.	EDM02.01	Formal documentation of stakeholder requirements and direction with regard to risk tolerance levels in IT investment policy	Formal documentation of risk function's contribution to business objectives	Internal

Risk-specific Activities (in Addition to COBIT 5 Activities)

1. Establish a method of demonstrating the value of risk management (including defining and collecting relevant data) to ensure the efficient use of assets.
2. Demonstrate the value of the risk function by highlighting the risk function's contribution to the business objectives.

Governance Practice	Risk-specific Inputs (in Addition to COBIT 5 Inputs)		Risk-specific Outputs (in Addition to COBIT 5 Outputs)	
	From	Description	Description	To
EDM02.03 Monitor value optimisation. Monitor the key goals and metrics to determine the extent to which the business is generating the expected value and benefits to the enterprise from IT-enabled investments and services. Identify significant issues and consider corrective actions.	EDM01.01	Decision-making model for IT risk	Feedback on delivery of risk initiatives	Internal
	EDM02.01	Formal documentation of stakeholder requirements and direction with regard to risk tolerance levels in IT investment policy		

Risk-specific Activities (in Addition to COBIT 5 Activities)

1. Monitor the value of risk initiatives and compare to stakeholder requirements that were set to ensure value delivery.
2. Use business-focussed methods of reporting on the added value of risk management initiatives.

Page intentionally left blank

EDM05 Ensure Stakeholder Transparency	Area: Governance Domain: Evaluate, Direct and Monitor

COBIT 5 Process Description
Ensure that enterprise IT performance and conformance measurement and reporting are transparent, with stakeholders approving the goals and metrics and the necessary remedial actions.

COBIT 5 Process Purpose Statement
Make sure that the communication to stakeholders is effective and timely and the basis for reporting is established to increase performance, identify areas for improvement, and confirm that IT-related objectives and strategies are in line with the enterprise's strategy.

EDM05 Risk-specific Process Goals and Metrics

Risk-specific Process Goals	Related Metrics
1. Risk reporting is established and is complete, timely and accurate.	• Percent of reports that are delivered on time • Percent of reports with validated reporting data
2. Stakeholders are informed of the current status of risk and risk management across the enterprise.	• Stakeholder satisfaction with the risk management reporting process (timely, complete, relevant, reliable, accurate, etc.) and frequency, based on surveys • Number of stakeholders that receive risk insights

EDM05 Risk-specific Process Practices, Inputs/Outputs and Activities

Governance Practice	Risk-specific Inputs (in Addition to COBIT 5 Inputs)		Risk-specific Outputs (in Addition to COBIT 5 Outputs)	
	From	Description	Description	To
EDM05.01 Evaluate stakeholder reporting requirements. Continually examine and make judgement on the current and future requirements for stakeholder communication and reporting, including both mandatory reporting requirements (e.g., regulatory) and communication to other stakeholders. Establish the principles for communication.	Outside *COBIT 5 for Risk*	Evaluation of enterprise reporting requirements	Evaluation of risk reporting requirements and communication channels	EDM05.03

Risk-specific Activities (in Addition to COBIT 5 Activities)

1. Determine the audience, including internal and external individuals or groups, for communication and reporting.
2. Examine and make a judgement on the current and future reporting requirements relating to IT risk within the enterprise (regulation, legislation, common law, contractual), including extent and frequency.
3. Identify the means and channels to communicate risk issues.

Governance Practice	Risk-specific Inputs (in Addition to COBIT 5 Inputs)		Risk-specific Outputs (in Addition to COBIT 5 Outputs)	
	From	Description	Description	To
EDM05.02 Direct stakeholder communication and reporting. Ensure the establishment of effective stakeholder communication and reporting, including mechanisms for ensuring the quality and completeness of information, oversight of mandatory reporting, and creating a communication strategy for stakeholders.	EDM01.01	Stakeholder requirements with regard to risk priorities and objectives	Summary of activities to enterprise risk committee	Internal

Risk-specific Activities (in Addition to COBIT 5 Activities)

1. Prioritise reporting on risk issues to stakeholders.
2. Produce for stakeholders regular risk and risk management status reports that include risk management activities, risk treatment against target dates, performance, achievements, risk profile, business benefits, 'hot topics' (e.g., cloud, consumer products), outstanding risk and capability gaps.

Governance Practice	Risk-specific Inputs (in Addition to COBIT 5 Inputs)		Risk-specific Outputs (in Addition to COBIT 5 Outputs)	
	From	Description	Description	To
EDM05.03 Monitor stakeholder communication. Monitor the effectiveness of stakeholder communication. Assess mechanisms for ensuring accuracy, reliability and effectiveness, and ascertain whether the requirements of different stakeholders are met.	EDM05.01	Evaluation of risk reporting requirements and communication channels	Risk monitoring and reporting	Internal

Risk-specific Activities (in Addition to COBIT 5 Activities)

1. Establish risk monitoring and reporting (e.g., using KPIs for risk and risk management that are based on metrics and measurements in the MEA domain).

Page intentionally left blank

APO02 Manage Strategy	Area: Management Domain: Align, Plan and Organise

COBIT 5 Process Description
Provide a holistic view of the current business and IT environment, the future direction, and the initiatives required to migrate to the desired future environment. Leverage enterprise architecture building blocks and components, including externally provided services and related capabilities to enable nimble, reliable and efficient response to strategic objectives.

COBIT 5 Process Purpose Statement
Align strategic IT plans with business objectives. Clearly communicate the objectives and associated accountabilities so they are understood by all, with the IT strategic options identified, structured and integrated with the business plans.

APO02 Risk-specific Process Goals and Metrics

Risk-specific Process Goals	Related Metrics
1. The risk function charter is defined and maintained.	• Stakeholder approval of the risk function charter
2. The risk function strategy is cost-effective, appropriate, realistic, achievable, enterprise-focused and balanced.	• Percent and number of initiatives for which a value metric (e.g., return on investment [ROI]) has been calculated • Enterprise stakeholder satisfaction survey feedback on the effectiveness of the risk management strategy
3. The risk function strategy is aligned with long- and short-term enterprise strategic goals and objectives.	• Percent of projects in the enterprise and IT project portfolios that involve risk management • Percent of IT projects that have risk requirements championed by business owners

APO02 Risk-specific Process Practices, Inputs/Outputs and Activities

Management Practice	Risk-specific Inputs (in Addition to COBIT 5 Inputs)		Risk-specific Outputs (in Addition to COBIT 5 Outputs)	
	From	Description	Description	To
APO02.01 Understand enterprise direction. Consider the current enterprise environment and business processes, as well as the enterprise strategy and future objectives. Consider also the external environment of the enterprise (industry drivers, relevant regulations, basis for competition).			Listing of potential risk function coverage gaps	APO02.02

Risk-specific Activities (in Addition to COBIT 5 Activities)

1. Understand how the risk function should support the overall enterprise objectives and protect stakeholder interests by taking into account the need to manage risk while meeting regulatory requirements and adding value to the enterprise.
2. Understand the current enterprise architecture and identify potential risk function coverage gaps.

Management Practice	Risk-specific Inputs (in Addition to COBIT 5 Inputs)		Risk-specific Outputs (in Addition to COBIT 5 Outputs)	
	From	Description	Description	To
APO02.02 Assess the current environment, capabilities and performance. Assess the performance of current internal business and IT capabilities and external IT services, and develop an understanding of the enterprise architecture in relation to IT. Identify issues currently being experienced and develop recommendations in areas that could benefit from improvement. Consider service provider differentiators and options and the financial impact and potential costs and benefits of using external services.	APO02.01	Listing of potential risk function coverage gaps	Risk function capabilities	APO02.03
			Baseline of the current business and IT environment	APO02.04

Risk-specific Activities (in Addition to COBIT 5 Activities)

1. Develop a baseline of the current business and IT environment, capabilities and services against which future requirements can be compared. This should include the relevant high-level detail of the current enterprise architecture (business, information, data, applications and technology domains), risk, business processes, IT processes and procedures, the IT organisation structure, external service provision, governance of IT, and enterprisewide IT-related skills and competencies.
2. Identify risks from current, potential and declining technologies.
3. Identify gaps between current business and IT capabilities and services and reference standards and best practices, competitor business and IT capabilities, and comparative benchmarks of best practice and emerging IT service provision.
4. Identify issues, strengths, opportunities and threats in the current environment, capabilities and services to understand current performance, and identify areas for improvement in terms of IT's contribution to enterprise objectives.

APO02 Risk-specific Process Practices, Inputs/Outputs and Activities *(cont.)*				
	Risk-specific Inputs (in Addition to COBIT 5 Inputs)		**Risk-specific Outputs (in Addition to COBIT 5 Outputs)**	
Management Practice	**From**	**Description**	**Description**	**To**
APO02.03 Define the target IT capabilities. Define the target business and IT capabilities and required IT services. This should be based on the understanding of the enterprise environment and requirements; the assessment of the current business process and IT environment and issues; and consideration of reference standards, best practices and validated emerging technologies or innovation proposals.	APO02.02	Risk function capabilities	Risk management requirements in target IT capabilities	APO02.04 APO02.05

Risk-specific Activities (in Addition to COBIT 5 Activities)
1. Ensure proper risk analysis is done when defining target IT capabilities. Identify threats from declining, current and newly acquired technology. 2. Agree on the impact of risk in changes of the enterprise architecture (business, information, data, applications and technology domains), business and IT processes and procedures, the IT organisation structure, IT service providers, governance of IT, and IT skills and competencies.

	Risk-specific Inputs (in Addition to COBIT 5 Inputs)		**Risk-specific Outputs (in Addition to COBIT 5 Outputs)**	
Management Practice	**From**	**Description**	**Description**	**To**
APO02.04 Conduct a gap analysis. Identify the gaps between the current and target environments and consider the alignment of assets (the capabilities that support services) with business outcomes to optimise investment in and utilisation of the internal and external asset base. Consider the critical success factors to support strategy execution.	APO02.02	Baseline of the current business and IT environment	Risk management coverage gaps to be closed	APO02.05
	APO02.03	Risk management requirements in target IT capabilities		
	Outside *COBIT 5 for Risk*	Regulation and compliance requirements for enterprise IT		

Risk-specific Activities (in Addition to COBIT 5 Activities)
1. Identify all gaps to be closed and changes required to realise the target environment, in light of the risk management processes, requirements and risk appetite of the enterprise. 2. Examine the current environment with respect to regulations and compliance requirements. 3. Where risk is identified and the decision is to accept the risk, then the risk acceptance process is to be adhered to.

	Risk-specific Inputs (in Addition to COBIT 5 Inputs)		**Risk-specific Outputs (in Addition to COBIT 5 Outputs)**	
Management Practice	**From**	**Description**	**Description**	**To**
APO02.05 Define the strategic plan and road map. Create a strategic plan that defines, in co-operation with relevant stakeholders, how IT- related goals will contribute to the enterprise's strategic goals. Include how IT will support IT- enabled investment programmes, business processes, IT services and IT assets. Direct IT to define the initiatives that will be required to close the gaps, the sourcing strategy and the measurements to be used to monitor achievement of goals, then prioritise the initiatives and combine them in a high-level road map.	APO02.03	Risk management requirements in target IT capabilities	Risk management strategy	APO02.06
	APO02.04	Risk management coverage gaps to be closed	Updated IT strategic plan and road map taking into account the risk management requirements	Internal

Risk-specific Activities (in Addition to COBIT 5 Activities)
1. Define the risk function strategy and align it with business strategies and the enterprise's overall objectives and risk appetite. 2. Ensure that the current IT strategic plan and road map take into account the risk requirements.

APO02 Risk-specific Process Practices, Inputs/Outputs and Activities *(cont.)*				
	Risk-specific Inputs (in Addition to COBIT 5 Inputs)		Risk-specific Outputs (in Addition to COBIT 5 Outputs)	
Management Practice	From	Description	Description	To
APO02.06 Communicate the IT strategy and direction. Create awareness and understanding of the business and IT objectives and direction, as captured in the IT strategy, through communication to appropriate stakeholders andusers throughout the enterprise.	APO02.05	Risk management strategy	Risk management plan	APO07.01 APO11.01
Risk-specific Activities (in Addition to COBIT 5 Activities)				
1. Develop the risk management plan based on the risk management strategy. 2. Gain approval from authorised stakeholders (e.g., CIO, executive management, board of directors) and communicate the risk management strategy and plan to all relevant stakeholders.				

Page intentionally left blank

APO06 Manage Budget and Costs	Area: Management Domain: Align, Plan and Organise

COBIT 5 Process Description
Manage the IT-related financial activities in both the business and IT functions, covering budget, cost and benefit management, and prioritisation of spending through the use of formal budgeting practices and a fair and equitable system of allocating costs to the enterprise. Consult stakeholders to identify and control the total costs and benefits within the context of the IT strategic and tactical plans, and initiate corrective action where needed.

COBIT 5 Process Purpose Statement
Foster partnership between IT and enterprise stakeholders to enable the effective and efficient use of IT-related resources and provide transparency and accountability of the cost and business value of solutions and services. Enable the enterprise to make informed decisions regarding the use of IT solutions and services.

APO06 Risk-specific Process Goals and Metrics

Risk-specific Process Goals	Related Metrics
1. Allocation of budget and costs for risk management is prioritised effectively.	• Percent of alignment between risk resources and high priority risk and control activities

APO06 Risk-specific Process Practices, Inputs/Outputs and Activities

Management Practice	Risk-specific Inputs (in Addition to COBIT 5 Inputs)		Risk-specific Outputs (in Addition to COBIT 5 Outputs)	
	From	Description	Description	To
APO06.01 Manage finance and accounting. Establish and maintain a method to account for all IT-related costs, investments and depreciation as an integral part of the enterprise financial systems and chart of accounts to manage the investments and costs of IT. Capture and allocate actual costs, analyse variances between forecasts and actual costs, and report using the enterprise's financial measurement systems.				

Risk-specific Activities (in Addition to COBIT 5 Activities)

Risk-specific guidance is not relevant for this practice. The generic COBIT 5 activities can be used as further guidance.

Management Practice	Risk-specific Inputs (in Addition to COBIT 5 Inputs)		Risk-specific Outputs (in Addition to COBIT 5 Outputs)	
	From	Description	Description	To
APO06.02 Prioritise resource allocation. Implement a decision-making process to prioritise the allocation of resources and rules for discretionary investments by individual business units. Include the potential use of external service providers and consider the buy, develop and rent options.				

Risk-specific Activities (in Addition to COBIT 5 Activities)

1. Establish a decision-making body for prioritising business and IT resources including the use of risk maps and the use of external service providers within the high-level budget allocations for IT-enabled programmes, IT services and IT assets as established by the strategic and tactical plans. Consider the options for buying or developing capitalised assets and services versus externally utilised assets and services on a pay-for-use basis.

Management Practice	Risk-specific Inputs (in Addition to COBIT 5 Inputs)		Risk-specific Outputs (in Addition to COBIT 5 Outputs)	
	From	Description	Description	To
APO06.03 Create and maintain budgets. Prepare a budget reflecting the investment priorities supporting strategic objectives based on the portfolio of IT-enabled programmes and IT services.			Risk function budget	Internal

Risk-specific Activities (in Addition to COBIT 5 Activities)

1. Develop a risk function budget.

AP006 Risk-specific Process Practices, Inputs/Outputs and Activities *(cont.)*				
	Risk-specific Inputs (in Addition to COBIT 5 Inputs)		**Risk-specific Outputs (in Addition to COBIT 5 Outputs)**	
Management Practice	**From**	**Description**	**Description**	**To**
AP006.04 Model and allocate costs. Establish and use an IT costing model based on the service definition, ensuring that allocation of costs for services is identifiable, measurable and predictable, to encourage the responsible use of resources including those provided by service providers. Regularly review and benchmark the appropriateness of the cost/chargeback model to maintain its relevance and appropriateness to the evolving business and IT activities.				
Risk-specific Activities (in Addition to COBIT 5 Activities)				
Risk-specific guidance is not relevant for this practice. The generic COBIT 5 activities can be used as further guidance.				
	Risk-specific Inputs (in Addition to COBIT 5 Inputs)		**Risk-specific Outputs (in Addition to COBIT 5 Outputs)**	
Management Practice	**From**	**Description**	**Description**	**To**
AP006.05 Manage costs. Implement a cost management process comparing actual costs to budgets. Costs should be monitored and reported and, in case of deviations, identified in a timely manner and their impact on enterprise processes and services assessed.				
Risk-specific Activities (in Addition to COBIT 5 Activities)				
Risk-specific guidance is not relevant for this practice. The generic COBIT 5 activities can be used as further guidance.				

APO07 Manage Human Resources	Area: Management Domain: Align, Plan and Organise

COBIT 5 Process Description
Provide a structured approach to ensure optimal structuring, placement, decision rights and skills of human resources. This includes communicating the defined roles and responsibilities, learning and growth plans, and performance expectations, supported with competent and motivated people.

COBIT 5 Process Purpose Statement
Optimise human resources capabilities to meet enterprise objectives.

APO07 Risk-specific Process Goals and Metrics

Risk-specific Process Goals	Related Metrics
1. HR capabilities and processes are aligned with the risk function requirements.	• Rate of turnover in risk function • Qualifications of staff in terms of certifications, education and years of experience

APO07 Risk-specific Process Practices, Inputs/Outputs and Activities

Management Practice	Risk-specific Inputs (in Addition to COBIT 5 Inputs)		Risk-specific Outputs (in Addition to COBIT 5 Outputs)	
	From	Description	Description	To
APO07.01 Maintain adequate and appropriate staffing. Evaluate staffing requirements on a regular basis or upon major changes to the enterprise or operational or IT environments to ensure that the enterprise has sufficient human resources to support enterprise goals and objectives. Staffing includes both internal and external resources.	APO02.06	Risk management plan	Risk function requirements for the staffing process	Internal

Risk-specific Activities (in Addition to COBIT 5 Activities)

1. Identify resource requirements for IT risk management at both the business and IT levels and in the context of competing business risk issues, resource limitations and objectives.
2. Allocate appropriate funds to fill gaps and position the enterprise to take advantage of opportunities.
3. Make risk/reward trade-offs in relation to organisational objectives.
4. Consider required skills (specify how the risk management skills of managers and staff will be developed and maintained), documented processes and procedures for IT risk management, information systems and databases for managing IT risk issues, budget and other resources for specific risk response activities, expectations from regulators and external auditors, etc.

Management Practice	Risk-specific Inputs (in Addition to COBIT 5 Inputs)		Risk-specific Outputs (in Addition to COBIT 5 Outputs)	
	From	Description	Description	To
APO07.02 Identify key IT personnel. Identify key IT personnel while minimising reliance on a single individual performing a critical job function through knowledge capture (documentation), knowledge sharing, succession planning and staff backup.				

Risk-specific Activities (in Addition to COBIT 5 Activities)

Risk-specific guidance is not relevant for this practice. The generic COBIT 5 activities can be used as further guidance.

AP007 Risk-specific Process Practices, Inputs/Outputs and Activities (cont.)

Management Practice	Risk-specific Inputs (in Addition to COBIT 5 Inputs)		Risk-specific Outputs (in Addition to COBIT 5 Outputs)	
	From	Description	Description	To
AP007.03 Maintain the skills and competencies of personnel. Define and manage the skills and competencies required of personnel. Regularly verify that personnel have the competencies to fulfil their roles on the basis of their education, training and/or experience, and verify that these competencies are being maintained, using qualification and certification programmes where appropriate. Provide employees with ongoing learning and opportunities to maintain their knowledge, skills and competencies at a level required to achieve enterprise goals.			Risk function training plan	AP007.04

Risk-specific Activities (in Addition to COBIT 5 Activities)

1. Provide professional development training and programmes on risk management.
2. Use certification to ensure a quality risk management professional skill set.
3. Establish appropriate enterprisewide education, training and awareness programmes for risk.

Management Practice	Risk-specific Inputs (in Addition to COBIT 5 Inputs)		Risk-specific Outputs (in Addition to COBIT 5 Outputs)	
	From	Description	Description	To
AP007.04 Evaluate employee job performance. Perform timely performance evaluations on a regular basis against individual objectives derived from the enterprise's goals, established standards, specific job responsibilities, and the skills and competency framework. Employees should receive coaching on performance and conduct whenever appropriate.	AP007.03	Risk function training plan	Risk function personnel evaluations	Internal
	Outside *COBIT 5 for Risk*	HR policy		

Risk-specific Activities (in Addition to COBIT 5 Activities)

1. Incorporate risk management criteria in the personnel evaluation process.

Management Practice	Risk-specific Inputs (in Addition to COBIT 5 Inputs)		Risk-specific Outputs (in Addition to COBIT 5 Outputs)	
	From	Description	Description	To
AP007.05 Plan and track the usage of IT and business human resources. Understand and track the current and future demand for business and IT human resources with responsibilities for enterprise IT. Identify shortfalls and provide input into sourcing plans, enterprise and IT recruitment processes sourcing plans, and business and IT recruitment processes.	Outside *COBIT 5 for Risk*	Process resource requirements, budget allocations, personnel lists, personnel skills	Resource performance tracking plan and indicators, resource allocation plan	Internal

Risk-specific Activities (in Addition to COBIT 5 Activities)

1. Manage allocation of risk management staff according to the risk management plan.

Management Practice	Risk-specific Inputs (in Addition to COBIT 5 Inputs)		Risk-specific Outputs (in Addition to COBIT 5 Outputs)	
	From	Description	Description	To
AP007.06 Manage contract staff. Ensure that consultants and contract personnel who support the enterprise with IT skills know and comply with the organisation's policies and meet agreed-on contractual requirements.				

Risk-specific Activities (in Addition to COBIT 5 Activities)

Risk-specific guidance is not relevant for this practice. The generic COBIT 5 activities can be used as further guidance.

APO08 Manage Relationships	Area: Management Domain: Align, Plan and Organise

COBIT 5 Process Description
Manage the relationship between the business and IT in a formalised and transparent way that ensures a focus on achieving a common and shared goal of successful enterprise outcomes in support of strategic goals and within the constraint of budgets and risk tolerance. Base the relationship on mutual trust, using open and understandable terms and common language and a willingness to take ownership and accountability for key decisions.

COBIT 5 Process Purpose Statement
Create improved outcomes, increased confidence, trust in IT and effective use of resources.

APO08 Risk-specific Process Goals and Metrics

Risk-specific Process Goals	Related Metrics
1. Co-ordination, communication and consultation are established between the risk function and other stakeholders.	• Percent of risk function representation in business committees • Number of direct communications from the risk function to various stakeholders
2. Stakeholders see the added value of risk management and recognise the risk function as a business enabler.	• Inclusion rate of risk management initiatives in investment proposals • Stakeholder satisfaction with risk management activities and results, measured through satisfaction surveys

APO08 Risk-specific Process Practices, Inputs/Outputs and Activities

Management Practice	Risk-specific Inputs (in Addition to COBIT 5 Inputs)		Risk-specific Outputs (in Addition to COBIT 5 Outputs)	
	From	Description	Description	To
APO08.01 Understand business expectations. Understand current business issues and objectives and business expectations for IT. Ensure that requirements are understood, managed and communicated, and their status agreed on and approved.	Outside *COBIT 5 for Risk*	Business goals and objectives	Understanding of business processes and expectations of the enterprise	APO08.02 APO08.03

Risk-specific Activities (in Addition to COBIT 5 Activities)

1. Understand the business and how IT risk enables/affects it.

Management Practice	Risk-specific Inputs (in Addition to COBIT 5 Inputs)		Risk-specific Outputs (in Addition to COBIT 5 Outputs)	
	From	Description	Description	To
APO08.02 Identify opportunities, risk and constraints for IT to enhance the business. Identify potential opportunities for IT to be an enabler of enhanced enterprise performance.	APO08.01	Understanding of business processes and expectations of the enterprise		

Risk-specific Activities (in Addition to COBIT 5 Activities)

1. Understand risk management trends and new technologies and how these can be applied innovatively to enhance business process performance.

Management Practice	Risk-specific Inputs (in Addition to COBIT 5 Inputs)		Risk-specific Outputs (in Addition to COBIT 5 Outputs)	
	From	Description	Description	To
APO08.03 Manage the business relationship. Manage the relationship with customers (business representatives). Ensure that relationship roles and responsibilities are defined and assigned, and communication is facilitated.	APO08.01	Understanding of business processes and expectations of the enterprise	Strategy to obtain stakeholder commitment	Internal

Risk-specific Activities (in Addition to COBIT 5 Activities)

1. Establish an approach for interacting with key business stakeholders.

APO08 Risk-specific Process Practices, Inputs/Outputs and Activities *(cont.)*				
	Risk-specific Inputs **(in Addition to COBIT 5 Inputs)**		**Risk-specific Outputs** **(in Addition to COBIT 5 Outputs)**	
Management Practice	**From**	**Description**	**Description**	**To**
APO08.04 Co-ordinate and communicate. Work with stakeholders and co-ordinate the end-to-end delivery of IT services and solutions provided to the business.	Outside *COBIT 5 for Risk*	Enterprise communication plan	Risk management communication strategy	Internal
Risk-specific Activities (in Addition to COBIT 5 Activities)				
1. Establish appropriate communication channels between the risk function and the business. 2. Establish the appropriate reporting and metrics regarding risk management.				
	Risk-specific Inputs **(in Addition to COBIT 5 Inputs)**		**Risk-specific Outputs** **(in Addition to COBIT 5 Outputs)**	
Management Practice	**From**	**Description**	**Description**	**To**
APO08.05 Provide input to the continual improvement of services. Continually improve and evolve IT-enabled services and service delivery to the enterprise to align with changing enterprise and technology requirements.			Integration of risk management in continual improvement process	Internal
Risk-specific Activities (in Addition to COBIT 5 Activities)				
1. Use risk analysis results as input for defining action plans and continuously improving the business.				

APO11 Manage Quality	Area: Management Domain: Align, Plan and Organise

COBIT 5 Process Description
Define and communicate quality requirements in all processes, procedures and the related enterprise outcomes, including controls, ongoing monitoring, and the use of proven practices and standards in continuous improvement and efficiency efforts.

COBIT 5 Process Purpose Statement
Ensure consistent delivery of solutions and services to meet the quality requirements of the enterprise and satisfy stakeholder needs.

APO11 Risk-specific Process Goals and Metrics

Risk-specific Process Goals	Related Metrics
1. Quality requirements for the risk function services are defined and implemented.	• Stakeholder satisfaction with risk management activities and results, measured through satisfaction surveys • Frequency of reporting (weekly, monthly, quarterly, annually) • Percent of risk staff with professional credentials • Number of continuing professional education (CPE) hours or hours of attendance at training or industry events

APO11 Risk-specific Process Practices, Inputs/Outputs and Activities

Management Practice	Risk-specific Inputs (in Addition to COBIT 5 Inputs)		Risk-specific Outputs (in Addition to COBIT 5 Outputs)	
	From	Description	Description	To
APO11.01 Establish a quality management system (QMS). Establish and maintain a QMS that provides a standard, formal and continuous approach to quality management for information, enabling technology and business processes that are aligned with business requirements and enterprise quality management.	APO02.06	Risk management plan	Relevant risk function best practices and standards	APO11.02 MEA01.01

Risk-specific Activities (in Addition to COBIT 5 Activities)

1. Determine risk function best practices.

Management Practice	Risk-specific Inputs (in Addition to COBIT 5 Inputs)		Risk-specific Outputs (in Addition to COBIT 5 Outputs)	
	From	Description	Description	To
APO11.02 Define and manage quality standards, practices and procedures. Identify and maintain requirements, standards, procedures and practices for key processes to guide the enterprise in meeting the intent of the agreed-on QMS. This should be in line with the IT control framework requirements. Consider certification for key processes, organisational units, products or services.	APO11.01	Relevant risk function best practices and standards	Risk function quality standards	APO11.03 APO11.04

Risk-specific Activities (in Addition to COBIT 5 Activities)

1. Align the risk management practices with the QMS.
2. Consider the benefits and cost of external quality reviews.

Management Practice	Risk-specific Inputs (in Addition to COBIT 5 Inputs)		Risk-specific Outputs (in Addition to COBIT 5 Outputs)	
	From	Description	Description	To
APO11.03 Focus quality management on customers. Focus quality management on customers by determining their requirements and ensuring alignment with the quality management practices.	APO11.02	Risk function quality standards		

Risk-specific Activities (in Addition to COBIT 5 Activities)

Risk-specific guidance is not relevant for this practice. The generic COBIT 5 activities can be used as further guidance.

AP011 Risk-specific Process Practices, Inputs/Outputs and Activities *(cont.)*				
	Risk-specific Inputs (in Addition to COBIT 5 Inputs)		**Risk-specific Outputs (in Addition to COBIT 5 Outputs)**	
Management Practice	**From**	**Description**	**Description**	**To**
AP011.04 Perform quality monitoring, control and reviews. Monitor the quality of processes and services on an ongoing basis as defined by the QMS. Define, plan and implement measurements to monitor customer satisfaction with quality as well as the value the QMS provides. The information gathered should be used by the process owners to improve quality.	AP011.02	Risk function quality standards	Risk function quality metrics implemented in line with best practices	AP011.05 MEA01.01

Risk-specific Activities (in Addition to COBIT 5 Activities)
1. Define risk management quality metrics to measure the achievement of risk management requirements and the efficient functioning of risk controls.   2. Monitor the risk management quality metrics.   3. Take corrective action to address quality issues in the risk function.

	Risk-specific Inputs (in Addition to COBIT 5 Inputs)		**Risk-specific Outputs (in Addition to COBIT 5 Outputs)**	
Management Practice	**From**	**Description**	**Description**	**To**
AP011.05 Integrate quality management into solutions for development and service delivery. Incorporate relevant quality management practices into the definition, monitoring, reporting and ongoing management of solutions development and service offerings.	AP011.04	Risk function quality metrics implemented in line with best practices	Documented root causes for risk management issues with quality metrics	Internal

Risk-specific Activities (in Addition to COBIT 5 Activities)
1. Identify, document and communicate root causes for risk management issues with quality metrics.   2. Apply corrective practices to remediate quality issues.

	Risk-specific Inputs (in Addition to COBIT 5 Inputs)		**Risk-specific Outputs (in Addition to COBIT 5 Outputs)**	
Management Practice	**From**	**Description**	**Description**	**To**
AP011.06 Maintain continuous improvement. Maintain and regularly communicate an overall quality plan that promotes continuous improvement. This should include the need for, and benefits of, continuous improvement. Collect and analyse data about the QMS, and improve its effectiveness. Correct non-conformities to prevent recurrence. Promote a culture of quality and continual improvement.				

Risk-specific Activities (in Addition to COBIT 5 Activities)
Risk-specific guidance is not relevant for this practice. The generic COBIT 5 activities can be used as further guidance

BAI08 Manage Knowledge	Area: Management Domain: Build, Acquire and Implement

COBIT 5 Process Description
Maintain the availability of relevant, current, validated and reliable knowledge to support all process activities and to facilitate decision making. Plan for the identification, gathering, organising, maintaining, use and retirement of knowledge.

COBIT 5 Process Purpose Statement
Provide the knowledge required to support all staff in their work activities and for informed decision making and enhanced productivity.

BAI08 Risk-specific Process Goals and Metrics

Risk-specific Process Goals	Related Metrics
1. Knowledge sharing is supported with the proper safeguards.	• Number of information leakage events

BAI08 Risk-specific Process Practices, Inputs/Outputs and Activities

Management Practice	Risk-specific Inputs (in Addition to COBIT 5 Inputs)		Risk-specific Outputs (in Addition to COBIT 5 Outputs)	
	From	Description	Description	To
BAI08.01 Nurture and facilitate a knowledge-sharing culture. Devise and implement a scheme to nurture and facilitate a knowledge-sharing culture.	Outside *COBIT 5 for Risk*	Information security training and awareness programme		

Risk-specific Activities (in Addition to COBIT 5 Activities)

1. Provide awareness training for the risk function relative to the sharing of information.

Management Practice	Risk-specific Inputs (in Addition to COBIT 5 Inputs)		Risk-specific Outputs (in Addition to COBIT 5 Outputs)	
	From	Description	Description	To
BAI08.02 Identify and classify sources of information. Identify, validate and classify diverse sources of internal and external information required to enable effective use and operation of business processes and IT services.			Classification of risk function information	BAI08.04

Risk-specific Activities (in Addition to COBIT 5 Activities)

1. Support the use and sharing of information within the risk function relative to its classification and sensitivity.
2. Develop a structure for how to classify risk management information.

Management Practice	Risk-specific Inputs (in Addition to COBIT 5 Inputs)		Risk-specific Outputs (in Addition to COBIT 5 Outputs)	
	From	Description	Description	To
BAI08.03 Organise and contextualise information into knowledge. Organise information based upon classification criteria. Identify and create meaningful relationships between information elements and enable use of information. Identify owners and define and implement levels of access to knowledge resources.			Published knowledge repositories	BAI08.04

Risk-specific Activities (in Addition to COBIT 5 Activities)

1. Map risk management roles to knowledge areas and ensure that proper access control is in place for relevant information.

Management Practice	Risk-specific Inputs (in Addition to COBIT 5 Inputs)		Risk-specific Outputs (in Addition to COBIT 5 Outputs)	
	From	Description	Description	To
BAI08.04 Use and share knowledge. Propagate available knowledge resources to relevant stakeholders and communicate how these resources can be used to address different needs (e.g., problem solving, learning, strategic planning and decision making).	BAI08.02	Classification of risk function information	Access control over risk management information	Internal
	BAI08.03	Published knowledge repositories		

Risk-specific Activities (in Addition to COBIT 5 Activities)

1. Ensure that the proper measures for data risk (e.g., loss, theft, corruption) are in place.
2. Implement access controls through the use of policies and processes to restrict unauthorised use and sharing of risk management information.

BAI08 Risk-specific Process Practices, Inputs/Outputs and Activities *(cont.)*				
Management Practice	**Risk-specific Inputs (in Addition to COBIT 5 Inputs)**		**Risk-specific Outputs (in Addition to COBIT 5 Outputs)**	
	From	**Description**	**Description**	**To**
BAI08.05 Evaluate and retire information. Measure the use and evaluate the currency and relevance of information. Retire obsolete information.			Updated rules for knowledge retirement and information disposal	Internal
Risk-specific Activities (in Addition to COBIT 5 Activities)				
1. Securely dispose of risk management information. 2. Send confirmation to information owner or information custodian about disposal of information.				

MEA01 Monitor, Evaluate and Assess Performance and Conformance	Area: Management Domain: Monitor, Evaluate and Assess

COBIT 5 Process Description
Collect, validate and evaluate business, IT and process goals and metrics. Monitor that processes are performing against agreed-on performance and conformance goals and metrics and provide reporting that is systematic and timely.

COBIT 5 Process Purpose Statement
Provide transparency of performance and conformance and drive achievement of goals.

MEA01 Risk-specific Process Goals and Metrics

Risk-specific Process Goals	Related Metrics
1. Performance of the risk function is monitored on an ongoing basis.	• Percent of business processes that meet defined risk management requirements • Percent of survey results measuring the satisfaction of the delivery of risk function services
2. Risk management and risk management practices comply with internal requirements.	• Percent of risk management practices that satisfy internal compliance requirements

MEA01 Risk-specific Process Practices, Inputs/Outputs and Activities

Management Practice	Risk-specific Inputs (in Addition to COBIT 5 Inputs)		Risk-specific Outputs (in Addition to COBIT 5 Outputs)	
	From	Description	Description	To
MEA01.01 Establish a monitoring approach. Engage with stakeholders to establish and maintain a monitoring approach to define the objectives, scope and method for measuring business solution and service delivery and contribution to enterprise objectives. Integrate this approach with the corporate performance management system.	APO11.01	Relevant risk function best practices and standards	Risk management monitoring process and procedure	MEA01.02
	APO11.04	Risk function quality metrics implemented in line with best practices		
	EDM01.01	Stakeholder requirements with regard to risk priorities and objectives		

Risk-specific Activities (in Addition to COBIT 5 Activities)

1. Identify and confirm risk management stakeholders.
2. Engage with stakeholders and communicate the risk management requirements and objectives for monitoring and reporting.
3. Establish the risk management monitoring process and procedure.
4. Align and continually maintain the risk management monitoring and evaluation approach with the IT and enterprise approaches.
5. Agree on a life cycle management and change control process for risk management monitoring and reporting.
6. Request, prioritise and allocate resources for monitoring risk management.

Management Practice	Risk-specific Inputs (in Addition to COBIT 5 Inputs)		Risk-specific Outputs (in Addition to COBIT 5 Outputs)	
	From	Description	Description	To
MEA01.02 Set performance and conformance targets. Work with the stakeholders to define, periodically review, update and approve performance and conformance targets within the performance measurement system.	MEA01.01	Risk management monitoring process and procedure	Agreed-on risk management metrics and targets	MEA01.04

Risk-specific Activities (in Addition to COBIT 5 Activities)

1. Define risk management performance targets consistent with overall IT performance standards.
2. Communicate risk management performance and conformance targets with key due diligence stakeholders.
3. Evaluate whether the risk management goals and metrics are adequate (i.e., specific, measurable, achievable, relevant and time-bound)

MEA01 Risk-specific Process Practices, Inputs/Outputs and Activities *(cont.)*				
	Risk-specific Inputs (in Addition to COBIT 5 Inputs)		**Risk-specific Outputs (in Addition to COBIT 5 Outputs)**	
Management Practice	**From**	**Description**	**Description**	**To**
MEA01.03 Collect and process performance and conformance data. Collect and process timely and accurate data aligned with enterprise approaches.	Outside *COBIT 5 for Risk*	Applicable regulations	Processed monitoring data	Internal
Risk-specific Activities (in Addition to COBIT 5 Activities)				
1. Collect and analyse performance and conformance data relating to risk management. 2. Assess the efficiency, appropriateness and integrity of collected data.				
	Risk-specific Inputs (in Addition to COBIT 5 Inputs)		**Risk-specific Outputs (in Addition to COBIT 5 Outputs)**	
Management Practice	**From**	**Description**	**Description**	**To**
MEA01.04 Analyse and report performance. Periodically review and report performance against targets, using a method that provides a succinct all-around view of IT performance and fits within the enterprise monitoring system.	MEA01.02	Agreed-on risk management metrics and targets	Risk management performance reports and corrective action plan updates	MEA01.05
Risk-specific Activities (in Addition to COBIT 5 Activities)				
1. Design, implement and agree on a range of risk management performance reports. 2. Compare the performance values to internal targets and benchmarks.				
	Risk-specific Inputs (in Addition to COBIT 5 Inputs)		**Risk-specific Outputs (in Addition to COBIT 5 Outputs)**	
Management Practice	**From**	**Description**	**Description**	**To**
MEA01.05 Ensure the implementation of corrective actions. Assist stakeholders in identifying, initiating and tracking corrective actions in order to address anomalies.	MEA01.04	Risk management performance reports and corrective action plan updates	Tracking process for corrective actions on risk management issues	Internal
	Outside *COBIT 5 for Risk*	Escalation guidelines		
Risk-specific Activities (in Addition to COBIT 5 Activities)				
1. Develop a tracking process for corrective actions on risk management issues.				

MEA02 Monitor, Evaluate and Assess the System of Internal Control	Area: Management Domain: Monitor, Evaluate and Assess

COBIT 5 Process Description
Continuously monitor and evaluate the control environment, including self-assessments and independent assurance reviews. Enable management to identify control deficiencies and inefficiencies and to initiate improvement actions. Plan, organise and maintain standards for internal control assessment and assurance activities.

COBIT 5 Process Purpose Statement
Obtain transparency for key stakeholders on the adequacy of the system of internal controls and thus provide trust in operations, confidence in the achievement of enterprise objectives and an adequate understanding of residual risk.

MEA02 Risk-specific Process Goals and Metrics

Risk-specific Process Goals	Related Metrics
1. Risk management controls are deployed and operating effectively.	• Percent of processes that satisfy risk management control requirements • Percent of controls in which risk management control requirements are met

MEA02 Risk-specific Process Practices, Inputs/Outputs and Activities

Management Practice	Risk-specific Inputs (in Addition to COBIT 5 Inputs)		Risk-specific Outputs (in Addition to COBIT 5 Outputs)	
	From	Description	Description	To
MEA02.01 Monitor internal controls. Continuously monitor, benchmark and improve the IT control environment and control framework to meet organisational objectives.	Outside *COBIT 5 for Risk*	Independent external audits	Defined risk management assurance scope and approach to assess internal controls	MEA02.03

Risk-specific Activities (in Addition to COBIT 5 Activities)

1. Provide independent assurance over IT risk management. Monitor IT risk action plans and obtain assurance on the performance of key IT risk management practices and whether IT risk is being managed in line with risk appetite and tolerance.
2. Ensure that control activities are in place and exceptions are promptly reported, followed up and analysed, and appropriate corrective actions are prioritised and implemented according to the risk management profile (e.g., classify certain exceptions as key risk and others as non-key risk).
3. Maintain the IT internal control system, considering ongoing changes in business and IT risk, the organisational control environment, relevant business and IT processes, and IT risk. If gaps exist, evaluate and recommend changes.

Management Practice	Risk-specific Inputs (in Addition to COBIT 5 Inputs)		Risk-specific Outputs (in Addition to COBIT 5 Outputs)	
	From	Description	Description	To
MEA02.02 Review business process controls effectiveness. Review the operation of controls, including a review of monitoring and test evidence, to ensure that controls within business processes operate effectively. Include activities to maintain evidence of the effective operation of controls through mechanisms such as periodic testing of controls, continuous controls monitoring, independent assessments, command and control centres, and network operations centres. This provides the business with the assurance of control effectiveness to meet requirements related to business, regulatory and social responsibilities.			Evidence of effectiveness of risk management controls	Internal

Risk-specific Activities (in Addition to COBIT 5 Activities)

1. Measure the effectiveness of risk management controls.

MEA02 Risk-specific Process Practices, Inputs/Outputs and Activities (cont.)

Management Practice	Risk-specific Inputs (in Addition to COBIT 5 Inputs)		Risk-specific Outputs (in Addition to COBIT 5 Outputs)	
	From	Description	Description	To
MEA02.03 Perform control self-assessments. Encourage management and process owners to take positive ownership of control improvement through a continuing programme of self-assessment to evaluate the completeness and effectiveness of management's control over processes, policies and contracts.	MEA02.01	Defined risk management assurance scope and approach to assess internal controls	Risk management assurance assessments	MEA02.04

Risk-specific Activities (in Addition to COBIT 5 Activities)

1. Perform risk management assurance assessments (independent and self-assessment) to identify control weaknesses.

Management Practice	Risk-specific Inputs (in Addition to COBIT 5 Inputs)		Risk-specific Outputs (in Addition to COBIT 5 Outputs)	
	From	Description	Description	To
MEA02.04 Identify and report control deficiencies. Identify control deficiencies and analyse and identify their underlying root causes. Escalate control deficiencies and report to stakeholders.	MEA02.03	Risk management assurance assessments	Assessment results and remedial actions	MEA02.08

Risk-specific Activities (in Addition to COBIT 5 Activities)

1. Review risk management reports for control deficiencies. Report and address noted deficiencies.

Management Practice	Risk-specific Inputs (in Addition to COBIT 5 Inputs)		Risk-specific Outputs (in Addition to COBIT 5 Outputs)	
	From	Description	Description	To
MEA02.05 Ensure that assurance providers are independent and qualified. Ensure that the entities performing assurance are independent from the function, groups or organisations in scope. The entities performing assurance should demonstrate an appropriate attitude and appearance, competence in the skills and knowledge necessary to perform assurance, and adherence to codes of ethics and professional standards.			Competence in skills and knowledge	Internal

Risk-specific Activities (in Addition to COBIT 5 Activities)

1. Establish competencies and qualifications for the assurance provider.

Management Practice	Risk-specific Inputs (in Addition to COBIT 5 Inputs)		Risk-specific Outputs (in Addition to COBIT 5 Outputs)	
	From	Description	Description	To
MEA02.06 Plan assurance initiatives. Plan assurance initiatives based on enterprise objectives and strategic priorities, inherent risk, resource constraints, and sufficient knowledge of the enterprise.	Outside *COBIT 5 for Risk*	Engagement plan	Updated engagement plan	MEA02.07

Risk-specific Activities (in Addition to COBIT 5 Activities)

1. Provide independent assurance over IT risk management. Agree to the objectives of the risk management review.

Management Practice	Risk-specific Inputs (in Addition to COBIT 5 Inputs)		Risk-specific Outputs (in Addition to COBIT 5 Outputs)	
	From	Description	Description	To
MEA02.07 Scope assurance initiatives. Define and agree with management on the scope of the assurance initiative, based on the assurance objectives.	MEA02.06	Updated engagement plan	Updated engagement plan	MEA02.08

Risk-specific Activities (in Addition to COBIT 5 Activities)

1. Define practices to validate control design and outcomes and determine whether the level of effectiveness supports acceptable risk (required by organisational or process risk assessment).
2. Where control effectiveness is not acceptable, define practices to identify residual risk (in preparation for reporting).

MEA02 Risk-specific Process Practices, Inputs/Outputs and Activities *(cont.)*				
	Risk-specific Inputs (in Addition to COBIT 5 Inputs)		**Risk-specific Outputs (in Addition to COBIT 5 Outputs)**	
Management Practice	**From**	**Description**	**Description**	**To**
MEA02.08 Execute assurance initiatives. Execute the planned assurance initiative. Report on identified findings. Provide positive assurance opinions, where appropriate, and recommendations for improvement relating to identified operational performance, external compliance and internal control system residual risk.	MEA02.04	Assessment results and remedial actions	Risk report and recommendations	Internal
	MEA02.07	Updated engagement plan		
Risk-specific Activities (in Addition to COBIT 5 Activities)				
1. Produce and issue signed-off risk management reports.				

Page intentionally left blank

MEA03 Monitor, Evaluate and Assess Compliance With External Requirements	Area: Management Domain: Monitor, Evaluate and Assess

COBIT 5 Process Description
Evaluate that IT processes and IT-supported business processes are compliant with laws, regulations and contractual requirements. Obtain assurance that the requirements have been identified and complied with and integrate IT compliance with overall enterprise compliance.

COBIT 5 Process Purpose Statement
Ensure that the enterprise is compliant with all applicable external requirements.

MEA03 Risk-specific Process Goals and Metrics

Risk-specific Process Goals	Related Metrics
1. Risk management practices conform to external compliance requirements.	• Percent of risk management practices that satisfy external compliance requirements
2. Monitoring is conducted for new or revised external requirements with an impact on risk management.	• Number or percent of projects initiated by the risk function to implement new external requirements

MEA03 Risk-specific Process Practices, Inputs/Outputs and Activities

Management Practice	Risk-specific Inputs (in Addition to COBIT 5 Inputs)		Risk-specific Outputs (in Addition to COBIT 5 Outputs)	
	From	Description	Description	To
MEA03.01 Identify external compliance requirements. On a continuous basis, identify and monitor for changes in local and international laws, regulations and other external requirements that must be complied with from an IT perspective.	Outside *COBIT 5 for Risk*	Risk management standards and regulations	External risk management compliance requirements	MEA03.02

Risk-specific Activities (in Addition to COBIT 5 Activities)

1. Establish arrangements for monitoring risk management compliance to external requirements.
2. Determine external compliance requirements to be met (including legal, regulatory, privacy and contractual).
3. Identify risk management compliance targets for external requirements.
4. Identify and communicate sources of risk management material to help meet external compliance requirements.

Management Practice	Risk-specific Inputs (in Addition to COBIT 5 Inputs)		Risk-specific Outputs (in Addition to COBIT 5 Outputs)	
	From	Description	Description	To
MEA03.02 Optimise response to external requirements. Review and adjust policies, principles, standards, procedures and methodologies to ensure that legal, regulatory and contractual requirements are addressed and communicated. Consider industry standards, codes of good practice, and best practice guidance for adoption and adaptation.	MEA03.01	External risk management compliance requirements	Updated external requirements	MEA03.03
	Outside *COBIT 5 for Risk*	Applicable regulations		

Risk-specific Activities (in Addition to COBIT 5 Activities)

1. Review and communicate external requirements to all relevant stakeholders.

Management Practice	Risk-specific Inputs (in Addition to COBIT 5 Inputs)		Risk-specific Outputs (in Addition to COBIT 5 Outputs)	
	From	Description	Description	To
MEA03.03 Confirm external compliance. Confirm compliance of policies, principles, standards, procedures and methodologies with legal, regulatory and contractual requirements.	MEA03.02	Updated external requirements	Risk management compliance report	Internal

Risk-specific Activities (in Addition to COBIT 5 Activities)

1. Collect and analyse compliance data relating to risk management.

MEA03 Risk-specific Process Practices, Inputs/Outputs and Activities *(cont.)*				
	Risk-specific Inputs (in Addition to COBIT 5 Inputs)		Risk-specific Outputs (in Addition to COBIT 5 Outputs)	
Management Practice	From	Description	Description	To
MEA03.04 Obtain assurance of external compliance. Obtain and report assurance of compliance and adherence with policies, principles, standards, procedures and methodologies. Confirm that corrective actions to address compliance gaps are closed in a timely manner.			Compliance assurance report	Internal
Risk-specific Activities (in Addition to COBIT 5 Activities)				
1. Gain evidence from the third parties.				

B.3. Enabler: Organisational Structures

This section contains detailed descriptions of the following organisational structures relevant for the risk function:
1. Enterprise risk management (ERM) committee, **figure 56**
2. Enterprise risk group, **figure 57**
3. Risk function. **figure 58**
4. Audit department, **figure 59**
5. Compliance department, **figure 60**

The detailed descriptions of these organisational structures are meant to be used as generic guidance or examples, and therefore do not take into account specific situations of enterprises. For example, in the case of a parent company with multiple subsidiaries, an ERM committee may or may not exist in every subsidiary. Span of control of these subsidiary organisations may be limited within this specific enterprise, not covering all group subsidiaries.

The organisational structures described are examples, not prescriptive. As a matter of fact, not every enterprise—depending on their nature, size, or any other contextual circumstance—may have implemented all of these structures.

Figures 56 through **60** describe the composition of the organisational structures by member roles and functions. The RACI chart lists a representative example of process practices where the organisational structure carries responsibilities. The level of approval within an enterprise is dependent on each enterprise's risk appetite or geographic regulatory requirements. The inputs that the structure needs to fulfil the mandate within the defined span of control are listed. Inputs can consist of particular items of information, policies or procedures, or decisions from other structures. The outputs that the structure generates by executing the mandate within the defined span of control are listed. Outputs can be decisions, particular pieces of information or policies/procedures.

Figure 56—Enterprise Risk Management (ERM) Committee	
Composition	
Role	**Description**
Risk function	The most senior official of the enterprise who is accountable for all aspects of risk management across the enterprise. An IT risk officer function may be established to oversee risk. In some enterprises the CEO will be charged with chairing the committee, per delegation by the board to oversee the day-to-day risk in the enterprise, when there is no specific CRO role.
CFO	The most senior official of the enterprise who is accountable for all aspects of financial management, including financial risk and controls and reliable and accurate accounts
CIO	The most senior official of the enterprise who is responsible for aligning IT and business strategies and accountable for planning, resourcing and managing the delivery of IT services and solutions to support enterprise objectives
COO	The most senior official of the enterprise who is accountable for the operations of the enterprise
Compliance department (representative)	The function in the enterprise responsible for guidance on legal, regulatory and contractual compliance
Audit department (representative)	The function in the enterprise responsible for conducting internal audit reviews and co-ordinating external audit
Privacy officer	An individual who is responsible for monitoring the risk and business impacts of privacy laws and for guiding and co-ordinating the implementation of policies and activities that will ensure that the privacy directives are met
CISO	The most senior officer of the enterprise who is accountable for the security of enterprise information in all forms
HR department representative	The most senior official of an enterprise who is accountable for planning and policies with respect to all human resources in that enterprise
Business owners (as appropriate for agenda under review)	A senior management individual accountable for the operation of a specific business unit or subsidiary. This includes key business line owners and heads of departments such as sales, marketing, human resources, manufacturing, etc.

Figure 56—Enterprise Risk Management (ERM) Committee *(cont.)*	
Mandate, Operating Principles, Span of Control and Authority Level	
Area	**Characteristic**
Mandate	Assist the board and the audit committee in supervising the ERM activities and advise the board with respect to the ERM framework.
Operating principles	• The ERM committee will meet on a regular basis (e.g., quarterly). More frequent meetings may be scheduled during specific initiatives or when issues need to be dealt with on a very urgent basis. • Reports to the board on a regular basis (e.g., quarterly) or as needed and makes recommendations on the activities covered by their mandate. • Minutes, including agenda, decisions made, attendance, action items and status reports, are approved by the committee (e.g., at the beginning of the next meeting) and retained in accordance with the enterprise retention policy. • Housekeeping rules: A chairperson will be named and the meetings run in accordance to the set of guiding principles such as Robert's Rules of Order.
Span of control	The ERM committee is servicing the entire legal entity for which the board is responsible
Authority level/decision rights	The ERM committee responsibilities include: • Direct the enterprise's risk management system, framework and methods(e.g., risk governance structure, assessment method, risk appetite) • Approve the ERM policy (under a delegation from the board) • Review the exposure • Oversee risk activities to ensure adherence to the board's defined risk appetite • Direct risk prioritisation and strategy Review the status of risk response and co-ordinate with business owners to allocate resources.
Delegation rights	The ERM committee has the right to delegate risk ownership to risk owners/subcommittees or business owners depending on the enterprise size and complexity.
Escalation path	All key issues and findings impacting the board's direction need to be escalated to the CEO and board.

RACI Chart	
Process Practice	**Level of Involvement (RACI)**
Steer the risk management system, framework and methods (e.g., risk governance structure, assessment method and risk appetite).	A
Review and approve the ERM policy.	R
Review the exposure and tolerance, aligned with the enterprise's risk appetite.	A
Institutionalise the enterprise's risk appetite.	A
Steer and approve risk prioritisation and strategy.	A
Review the status of risk closure and co-ordinate with business owners the appropriate allocation of resources.	A
Report the exposure to the board.	A

Figure 56—Enterprise Risk Management (ERM) Committee *(cont.)*		
Inputs/Outputs		
Input	**Type**	**From**
Risk KRIs, KPIs and key goal indicators (KGIs)	Information	Performance reporting
Incident reports	Information	Incident process
Business strategy (e.g., emerging technologies)	Information	Strategy process
Policies	Decision	Enterprise governance
Audit reports or other reviews	Information	• Audit reviews (internal and external) • Security assessments and tests, business continuity and resiliency service (BCRS) tests, etc.
Risk report (current and mitigation status)	Decision	Risk management
Risk register	Information	Risk management
Regulations	Information	Legal and compliance
Threat intelligence	Information	Threat intelligence providers (internal and external)
Output	**Type**	**To**
Risk tolerance	Information	Risk Management
Meeting minutes externals as appropriate	Decision	Risk management
Risk management strategy	Decision	• Board • Business owners
Risk mitigation actions	Decision	• Board • Business owners • IT process owners
Policy (change control)	Decision	Communications

Figure 57—Enterprise Risk Group	
Composition	
Role	**Description**
Risk function	The most senior official of the enterprise who is accountable for all aspects of risk management across the enterprise. This role requires risk-specific technical expertise to govern the risk, direct capabilities to manage the risk management group, communicate and influence capabilities to effectively interact with the stakeholders.
Risk managers	This role requires risk specific expertise to establish, manage and sustain risk management processes. Strong interpersonal capabilities are required to engage stakeholders as risk owners to undertake risk processes (such as risk identification, risk rating, assessment etc.). Effective communication skills are required to represent risk analysis results to the CRO and influence the design and implementation of controls.
Risk analysts	This role requires expertise in the breakdown of complex risk data, analysis of interactions/dependencies and effective communication of findings and trends. It also requires knowledge and practical experience with risk frameworks, methodologies, commonly used risk standards and risk best practices.
Technical experts (e.g., IT security, Oracle® expert, business process expert)	This role should have the technical expertise necessary to analyse the areas of risk and in terms of their vulnerabilities and threats, not only to understand how events can lead to incidents (risk scenarios), but also to provide information on root causes of certain incidents and suggest controls.
Mandate, Operating Principles, Span of Control and Authority Level	
Area	**Characteristic**
Mandate	The enterprise risk group is established to consider risk in more detail and advise the ERM committee.
Operating principles	Via the CRO, it will report to the CEO and ERM committee on a regular basis (e.g., quarterly) on all issues in the span of this mandate. More frequent meetings may be scheduled when urgent issues are present.
Span of control	The enterprise risk group is servicing the entire legal entity for which the board is responsible.
Authority level/decision rights	Manage and execute the ERM programme: • Identify risk • Analyse risk to determine impact • Assess the impact of the risk to the aggregated risk profile • Analyse, together with business process owners, the risk (and propose an appropriate response) Monitor and review the risk environment.
Delegation rights	The enterprise risk group shares responsibility on process risk with the business process owners.
Escalation path	All key issues and findings impacting the board's direction need to be escalated to the ERM committee, CEO and board.

Figure 57—Enterprise Risk Group *(cont.)*	
RACI Chart	
Process Practice	**Level of Involvement (RACI)**
Research, define and document the enterprise's risk management requirements.	R
Define, manage and continuously improve the ERM system, method, framework and tools to help ensure consistent and effective risk acceptance decisions.	R
Develop the ERM strategy and implement the plan.	R
Develop risk management policy, procedures and other documents/templates.	R
Implement and manage the risk management process (APO12) and the supporting tools and knowledge repositories.	R
Collect and communicate the threats and vulnerability intelligence.	R
Provide subject matter expertise to identify current risk and projected risk associated with changes in the business strategy.	R
Provide a system to collect, categorise and report the enterprise exposure to risk while aligning to the business strategy and managing risk cost-effectively.	R
Establish a communications strategy to promote collaboration, understanding of risk ownership and the desired behaviours with stakeholders across the enterprise.	R

Inputs/Outputs		
Input	**Type**	**From**
Risk KRIs, KPIs and KGIs	Information	Performance reporting
Risk appetite and tolerance	Decision	ERM committee
ERM committee meeting minutes	Decision	ERM committee
Business strategy (e.g., emerging technologies)	Information	Strategy process
Policies	Decision	Enterprise governance
Audit reports or other reviews	Information	• Audit reviews (internal and external) • Security assessments and tests, BCRS tests, etc.
Regulations	Information	Legal and compliance
Threat intelligence	Information	Threat intelligence providers (internal and external)
Output	**Type**	**To**
Risk report (current and mitigation status)	Decision	ERM committee
Risk register	Information	• ERM committee • Business process owners
Risk mitigation actions	Decision	• ERM committee • Business process owners • IT process owners

Figure 58—Risk Function	
Mandate, Operating Principles, Span of Control and Authority Level	
Area	**Characteristic**
Mandate	Overall responsibility for the development and implementation of the ERM programme
Operating principles	• Ensures a holistic approach to risk management • Sets the parameters of the risk framework, maintains the risk profile and reports on significant risk issues to the CEO and ERM committee on a regular basis
Span of control	The entire enterprise risk environment
Authority level/decision rights	Directs and manages the ERM programme
Delegation rights	Assigns risk ownership and responsibilities
Escalation path	Escalates to the CEO and ERM committee

Figure 58—Risk Function *(cont.)*	
RACI Chart	
Process Practice	**Level of Involvement (RACI)**
Research, define and document the enterprises risk management requirements.	R
Define, manage, and continuously improve the ERM system, method, framework and tools to help ensure consistent and effective risk acceptance decisions.	R
Develop the ERM strategy and implement the plan	R
Develop risk management policy, procedures and other documents/templates.	A
Implement and manage the risk management process (APO12) and the supporting tools and knowledge repositories.	R
Collect and communicate threats and vulnerability intelligence.	R
Provide subject matter expertise to identify current risk and projected ones associated with changes in the business strategy.	A
Provide a system to measure and report the enterprise loss exposure while aligning to the business strategy and managing risk response effectively.	A/R
Establish a communications strategy to promote collaboration, understanding of risk ownership and the desired behaviours with stakeholders across the enterprise.	A/R
Define a risk management action portfolio and manage opportunities to reduce risk to an acceptable level.	A

Figure 59—Audit Department	
Mandate, Operating Principles, Span of Control and Authority Level	
Area	**Characteristic**
Mandate	Responsible for testing, examination and reporting of control conditions within the enterprise. The mandate will be determined in the audit charter. The department will define and execute an audit plan usually based on business risk considerations.
Operating principles	Independence from operational responsibilities. Auditors must ensure their independence is unquestionable when performing their audit duties, but, within this limitation, may give advice e.g., on business-related risk, business impact analysis (BIA), etc.
Span of control	Span of control is enterprisewide. The audit department is servicing the entire legal entity for which the board is responsible.
Authority level/decision rights	• Selecting audit processes and tools • Establish risk ratings for audit findings Direct reporting to the board
Delegation rights	Able to delegate some testing activities to other parts of the organisation or external subject matter experts provided that they can assure the quality and reliability of the tests.
Escalation path	Audit committee, the board and/or CEO
RACI Chart	
Process Practice	**Level of Involvement (RACI)**
Develop and execute of the enterprise audit plan in alignment with the enterprise strategy.	A/R
Plan and perform audits.	A/R
Communicate findings and validate resulting remediation plans.	A/R
Set audit finding severity.	A/R
Report enterprise control state to the board.	A/R
Submit audit plan for approval by the board.	A/R
Engage with stakeholders.	A/R

Figure 60—Compliance Department	
Mandate, Operating Principles, Span of Control and Authority Level	
Area	**Characteristic**
Mandate	To monitor and report on the state of compliance within the enterprise related to relevant laws and regulations.
Operating principles	Ensure that compliance requirements are understood, communicated and adhered to, and areas of non-compliance are remediated.
Span of control	Enterprisewide
Authority level/decision rights	• Set up a compliance program. • Recommend course of action to appropriate decision makers. • Assign ownership of compliance activities.
Delegation rights	N/A
Escalation path	General counsel, CEO, ERM committee and/or board.
RACI Chart	
Process Practice	**Level of Involvement (RACI)**
Monitor relevant laws and regulations.	A/R
Interpret and communicate requirements from relevant laws and regulations to affected stakeholders.	A/R
Facilitate compliance activities.	A/R
Collaborate with regulatory bodies.	A/R
Facilitate regulatory examinations.	A/R
Perform internal compliance audits.	A/R
Communicate compliance examination results to appropriate stakeholders.	A/R
Make recommendations for achieving and maintaining compliance.	A/R
Understand the impact of non-compliance on the organisation.	A/R

B.4. Enabler: Culture, Ethics and Behaviour

B.4.1 Influencing Behaviours

This section describes how the behaviours described in section 2A, chapter 6 can be influenced by leadership at different levels of the enterprise, i.e., at executive management level, at management of the risk function level and at the practitioner level.

Behaviour can be influenced through:
• The use of communication, enforcement and rules
• Incentives and rewards
• Raising awareness

Influencing Behaviour Through Communication, Enforcement and Rules
Leadership uses communication, enforcement and rules to influence behaviours in an enterprise. Communication is always essential to influence any kind of behaviour. Development and implementation of a risk-aware culture is dependent on the degree of importance of that aspect of a culture. Policies and processes can be used to enforce internal action where regulations are mandated.

Following are some examples of how executive management can use communication, enforcement and rules to influence and promote desirable behaviour:
• Tone at the top—management should lead by example as staff take cues for acceptable behaviour.
• Risk considerations are embedded in business planning.
• Organisational policies clearly identify the required compliance requirements and ensure that they are enforced.
• Senior management reinforces transparency by cascading appropriate communication throughout the enterprise.
• Zero-tolerance approach for non-ethical behaviours has clear consequences for misbehaviour.
• The CEO communicates to business unit heads to maintain a positive relationship with the assurance function.
• The CEO communicates to business unit heads the need to implement recommendations from risk analysis and propagates the use of root cause analysis.
• Top management approves and communicates a documented security policy based on which procedures, practices, standards and guidelines are prepared and shares with all relevant employees.
• Top management encourages active participation in enterprise initiatives.
• Top management establishes a speak-up process to promote and support whistle-blowing.
• Regulatory requirements as relevant to ethics are identified and communicated to all employees and are included in the code of ethics.
• Training budget, tools and guidance to support training efforts are provided.
• A communication plan is set up and executed to assist the enterprise change management when a new project, procedure, programme, etc. is launched.
• Within an enterprise that has established formal rules, informal structures can be established through positive communication. These informal structures give an extended reach to people who are rarely in contact with each other. This creates a positive synergy in the daily interaction between people.
• The right signals can produce positive effects, wrong signals can lead to undesirable outcomes. A positive communication can give extraordinary results in working teams.

Influencing Behaviour Through Incentives and Rewards
Management influences behaviour through measures designed to provide positive reinforcement for desired conduct and negative reinforcement for conduct it wishes to discourage. An absence of rewards inhibits adoption of a risk-aware culture. Business management needs to know that risk-aware behaviour will be rewarded; this, in turn, means that executive management must make its intentions clear by encouraging the implementation of appropriate risk responses and promoting attitudes that constitute a culture of risk awareness.

The following incentives and rewards may be used by the various management levels to influence behaviour:
• Management performance is aligned to risk management criteria and supports a risk-aware culture.
• Bonuses are paid on executive's conformance to risk practice and expected outcomes; sanctions and fines are imposed when demonstrating opposite behaviour.
• Engagement with risk is a recognised requirement in evaluation of management performance.
• Management has a policy of recognising and rewarding employees who obtain specific certification in the area of IT governance, assurance, security and control.
• A structured and visible career progression plan is set up. In doing so, employees will know that if they do the right things or add value, they can go up the ladder of their career.
• Management encourages practitioners among the staff:
  – Active participation is considered as a key attribute in the assessment of individual performance.
  – Ethical behaviour is seen a core requirement in performance evaluation.
• Public recognition is a highly valued reward to increase visibility and relationships with managers and peers.
• Financial recognition is commensurate with the level of value:
  – Be part of the whole.
  – Your role is considered an essential part of the machinery.
  – Personal performance goals, objectives, rewards and sanctions are formalized.

Influencing Behaviour Through Raising Awareness
Awareness programmes have their place, but they are insufficient by themselves. More than just being aware of information risk, people need to be educated about risk and their role in it. The different management levels in an enterprise can raise awareness through the following means.

Executive management can raise awareness by:
• Visibly supporting the development, execution and completion of training programmes to ensure a consistent risk-aware culture
• Regularly communicating to staff the need for risk management throughout the organisation to enforce their awareness of the risk environment as the basis for decision making

Risk management can raise awareness by:
• Running risk awareness workshops with key stakeholders to communicate key risk impacts and probabilities
• Creating awareness programmes that form the basis of what should be done once the risk materialises. These responses are industry- and organisational-specific, like the mitigation or action plans to be taken in the case of risk materialising.

Risk practitioners can raise awareness with:
• Case studies that are used to highlight the business impact of failures in risk awareness or compliance
• Enterprise literature/portal that highlights individuals who champion the desired values

B.5. Enabler: Information

This section provides details regarding the use and optimisation of risk-related information items, based on the introduction of the Information enabler in section 2A, chapter 6.

The information items presented include:
- **Figure 61—Risk Profile**
- **Figure 63—Risk Communication Plan**
- **Figure 64—Risk Report**
- **Figure 65—Risk Awareness Programme**
- **Figure 66—Risk Map**
- **Figure 67—Risk Universe, Appetite and Tolerance**
- **Figure 69—Key Risk Indicators**
- **Figure 71—Emerging Risk Issues and Factors**
- **Figure 72—Risk Taxonomy**
- **Figure 73—Business Impact Analysis**
- **Figure 74—Risk Event**
- **Figure 75—Risk and Control Activity Matrix**
- **Figure 76—Risk Assessment**

The information items are discussed in detail, and each dimension of the Information enabler model is elaborated on. This allows the user to consider:
- All goals or quality criteria for the information item that should be achieved, allowing development of appropriate applications and procedures
- The life cycle of the information item
- The stakeholders that should be included in the information life cycle

Figure 61—Risk Profile

A risk profile is a description of the overall (identified) risk to which the enterprise is exposed. A risk profile consists of:
- **Risk register**
  - **Risk scenarios**
  - **Risk analysis**
- **Risk action plan**
- **Loss events (historical and current)**
- **Risk factors**
- **Independent assessment findings**

	Life Cycle Stage	Internal Stakeholder	External Stakeholder	Description/Stake
Life Cycle and Stakeholders	**Information planning**	ERM committee, board	External audit, regulator	• Internal stakeholders: Initiate and drive the implementation and appoint a CRO. Have adequate information on the exposure. • External stakeholders: To have comfort on the risk management capabilities
	Information design	Risk function, compliance, CIO, CISO, business process owners, internal audit		• CRO: To obtain information from the other roles in order to provide the overview for the governance bodies • CIO: To be able to develop an adequate information system • Other roles: To be able to provide relevant information and to ensure completeness/adequacy
	Information build/acquire	Risk function, internal audit		• CRO: Provides functional requirements and consults others. • Internal audit: Provides quality assurance services on the implementation.
	Information use/operate: store, share, use	Board, ERM committee, business executive, CIO, risk function, CISO, business process owners, compliance, internal audit	External audit, regulator	• Business process owners, business executives and CIO: To efficiently provide relevant information • Board and ERM committee: To receive relevant information and to enable decision making • Internal audit, external audit and regulator: Receive relevant information • CRO: Oversees the caption, processing and interpretation of information
	Information monitor	Board, ERM committee, risk function, internal audit	External audit	• CRO: Ongoing monitoring on adequacy, completeness and accuracy of information; semi-annual assessment of performance (MEA01) and controls (MEA02) to maintain the information • Internal audit: Annual validation of format and level of contents
	Information dispose	Risk function		• CRO: According to data retention policy, to ensure confidentiality of information and to reduce the amount of information

Figure 61—Risk Profile *(cont.)*				
	Quality Subdimension and Goals	**Description—The extent to which information is…**	**Relevance**	**Goal**
Goals — Intrinsic	**Accuracy**	correct and reliable	High	Source information needs to be accurate (confirm through audit) and needs to be aggregated in the risk management application according to fixed rules.
	Objectivity	unbiased, unprejudiced and impartial	High	Information is based on verifiable facts and substantiations, using the common risk view established throughout the enterprise.
	Believability	regarded as true and credible	Medium	Reporting is fully trusted.
	Reputation	regarded as coming from a true and credible source	Medium	Source information is collected from competent and recognised sources.
Contextual and Representational	**Relevancy**	applicable and helpful for the task at hand	High	The risk profile is structured as defined and the recipient confirms the relevancy of information provided.
	Completeness	not missing and is of sufficient depth and breadth for the task at hand	High	The risk profile covers the full enterprise scope and the full risk register. However, not all information might be available and assumptions need to be made and substantiated.
	Currency	sufficiently up to date for the task at hand	Low	The need for currency of the risk profile is driven by the frequency and impact of alterations and depending on the component.
	Amount of information	is appropriate in volume for the task at hand	High	The volume of information is appropriate to the recipient's needs and shall be defined during the design.
	Concise representation	compactly represented	Medium	The risk profile is concisely represented; this is obtained by aggregating data for the entire enterprise and by retaining only individual cases from a predefined threshold onwards.
	Consistent representation	presented in the same format	Low	The risk profile is always presented according to an agreed-on template. However, single scenarios might be analysed differently when agreed on.
	Interpretability	in appropriate languages, symbols and units, and the definitions are clear	High	To ease decision making, the information 'sweet spot' can be identified and focused on.
	Understandability	easily comprehended	High	In order to make informed decisions, the risk profile should be understood by many stakeholders.
	Manipulation	easy to manipulate and apply to different tasks	Low	Scenarios can be modified and simulated.
Security	**Availability**	available when required, or easily and quickly retrievable	Medium	The risk profile is at all times available to its stakeholders; a temporary unavailability is acceptable in case of an incident.
	Restricted access	restricted appropriately to authorised parties	High	Access to the risk profile is determined by the risk function, and is restricted as follows: • Write access: Risk function (based on input from contributors) • Read access: All other stakeholders

Figure 61—Risk Profile *(cont.)*			
	Attribute	**Description**	**Value**
Good Practice	**Physical**	Information carrier/media	The information carrier for the risk profile can be an electronic or printed document or an information system (e.g., dashboard).
	Empiric	Information access channel	The risk profile is accessible through the ERM portal or printed at specific locations.
	Syntactic	Code/language	The risk profile contains the following subparts: • Risk register (results of risk analysis), which consists of a list of risk scenarios and their associated estimates for impact and frequency (risk map); both current and the previous risk map will be included. • Risk action plan, including action item, status, responsible, deadline, etc. • Loss data related to events occurring over the last reporting period(s) • Risk factors, including both contextual risk factors and capability-related risk factors (vulnerabilities) Result of independent assessments (e.g., audit findings, self-assessments)
	Semantic	Information type	Structured document based on a template and/or an online dashboard with drill-down functionality.
		Information currency	The risk profile contains historical, current and forward looking data.
		Information level	The risk profile aggregates data over the entire enterprise, representing only major risk over a defined threshold and with significant changes to previous periods.
	Pragmatic	Retention period	The risk profile is to be retained for as long as the data/information over which it reports risk needs to be retained. Updates to the risk register should be logged and retained as defined in legal requirements, use the information as evidence or the need to obtain independent assurance.
		Information status	The current instance is operational, older ones are historical data.
		Novelty	The risk profile combines several other sources of data that make up a new instance, hence it is novel data. It is updated regularly (e.g., on a monthly basis).
		Contingency	The risk profile relies on the following information being available and understood by the user: • Risk appetite of the enterprise • Risk factors that apply to the enterprise • Risk taxonomy in use in the enterprise
	Social	Context	The risk profile is primarily meaningful and to be used in a context of ERM, but could also be used in other circumstances (e.g., during a merger).

Figure 61—Risk Profile *(cont.)*	
Link to Other Enablers	
Processes	The risk profile is an <u>output</u> from the management practices: • APO12.03 Maintain a risk profile. • APO12.04 Articulate risk. The risk register is an <u>input</u> for the management practices: • EDM03.02 Direct risk management. • EDM05.02 Direct stakeholder communication and reporting. • APO02.02 Assess the current environment, capabilities and performance. • MEA02.08 Execute assurance initiatives. It is used in the following management practices: • EDM03.03 Monitor risk management. • APO12.06 Respond to risk. It is mentioned in the goal of the process APO12 *Manage Risk*: • A current and complete risk profile exists. And measured by the following metrics: • Percent of key business processes included in the risk profile • Completeness of attributes and values in the risk profile
Organisational Structures	Under the accountability of the CRO, the following roles are responsible for providing/producing the information: • Business process owners • CIO (and IT staff members) • CISO
Infrastructure, Applications and Services	The risk profile is produced by a risk management application or manually maintained by the CRO.
People, Skills and Competencies	The generation of the risk profile requires an understanding of risk management principles and skills. The provision of information requires subject-related expertise and the presentation of information should not require risk management skills but enable governance bodies to steer risk management and to take decisions.
Culture, Ethics and Behaviour	The availability of the risk profile supports the transparency of risk as well as trends and a risk-aware culture.
Principles, Policies and Frameworks	Related principles: • Connect to enterprise objectives • Align with ERM • Balance cost/benefit of IT risk • Consistent approach

One important component of the risk profile is the risk register. **Figure 62** contains a sample template for a risk register entry.

Figure 62—Template Risk Register Entry					
Part I—Summary Data					
Risk statement					
Risk owner					
Date of last risk assessment					
Due date for update of risk assessment					
Risk category	☐ STRATEGIC (IT Benefit/Value Enablement)		☐ PROJECT DELIVERY (IT Programme and Project Delivery)		☐ OPERATIONAL (IT Operations and Service Delivery)
Risk classification (copied from risk analysis results)	☐ LOW	☐ MEDIUM		☐ HIGH	☐ VERY HIGH
Risk response	☐ ACCEPT	☐ TRANSFER		☐ MITIGATE	☐ AVOID
Part II—Risk Description					
Title					
High-level scenario (from list of sample high-level scenarios)					
Detailed scenario description—Scenario components	**Actor**				
	Threat Type				
	Event				
	Asset/Resource				
	Timing				
Other scenario information					
Part III—Risk Analysis Results					
Frequency of scenario (number of times per year)	0	1	2	3	4	5
	$N \le 0{,}01$ ☐	$0{,}01 < N \le 0{,}1$ ☐	$0{,}1 < N \le 1$ ☐	$1 < N \le 10$ ☐	$10 < N \le 100$ ☐	$100 < N$ ☐
Comments on frequency					
Impact of scenario on business	0	1	2	3	4	5
1. Productivity	Revenue Loss Over One Year					
Impact rating	$I \le 0{,}1\%$ ☐	$0{,}1\% < I \le 1\%$ ☐	$1\% < I \le 3\%$ ☐	$3\% < I \le 5\%$ ☐	$5\% < I \le 10\%$ ☐	$10\% < I$ ☐
Detailed description of impact					
2. Cost of response	Expenses Associated With Managing the Loss Event					
Impact rating	$I \le 10k\$$ ☐	$10K\$ < I \le 100K\$$ ☐	$100K\$ < I \le 1M\$$ ☐	$1M\$\% < I \le 10M\$$ ☐	$10M\$ < I \le 100M\$$ ☐	$100M\$ < I$ ☐
Detailed description of impact					
3. Competitive advantage	Drop-in Customer Satisfaction Ratings					
Impact rating	$I \le 0{,}5$ ☐	$0{,}5 < I \le 1$ ☐	$1 < I \le 1{,}5$ ☐	$1{,}5 < I \le 2$ ☐	$2 < I \le 2{,}5$ ☐	$2{,}5 < I$ ☐
Detailed description of impact					
4. Legal	Regulatory Compliance—Fines					
Impact rating	None ☐	$< 1M\$$ ☐	$< 10M\$$ ☐	$< 100M\$$ ☐	$< 1B\$$ ☐	$> 1B\$$ ☐
Detailed description of impact					
Overall impact rating (average of four impact ratings)					
Overall rating of risk (obtained by combining frequency and impact ratings on risk map)	☐ LOW	☐ MEDIUM		☐ HIGH	☐ VERY HIGH

Figure 62—Template Risk Register Entry *(cont.)*				
Part III—Risk Response				
Risk response for this risk	☐ACCEPT	☐TRANSFER	☐MITIGATE	☐AVOID

Justification				

Detailed description of response (NOT in case of ACCEPT)	**Response Action**		**Completed**	**Action Plan**
	1.		☐	☐
	2.		☐	☐
	3.		☐	☐
	4.		☐	☐
	5.		☐	☐
	6.		☐	☐
Overall status of risk action plan				
Major issues with risk action plan				
Overall status of completed responses				
Major issues with completed responses				

Part IV—Risk Indicators	
Key risk indicators for this risk	1. 2. 3. 4.

Figure 63—Risk Communication Plan				
A risk communication plan defines frequency, types and recipients of information about risk. The main purpose of the plan is to reduce the overload of non-relevant information (avoiding the likelihood of 'risk noise').				

	Life Cycle Stage	**Internal Stakeholder**	**External Stakeholder**	**Description/Stake**
Life Cycle and Stakeholders	**Information planning**	ERM committee, audit committee		• Ensure that a risk communication plan exists and approve it. • Ensure that risk is communicated efficiently and timely.
	Information design	Risk function		• Outline the different aspects to be included in the risk communication plan (e.g., frequencies/types/recipient). • Ensure that risk is communicated efficiently and in a timely manner.
	Information build/acquire	Risk function, business process owners/CIO		• Detail the outlined aspects of the risk communication plan. • Ensure the risk communication plan includes key requirements for risk communication.
	Information use/operate: store, share, use	Board, executive management, CIO, risk function, business process owners, compliance, internal audit	External audit, regulator	Effective utilisation of the risk communication plan by the risk function ensures availability of correct and concise information about risk to the other stakeholders.
	Information monitor	Board, risk committee, audit committee, risk function	External audit	• Timely action is taken on the status of risk and risk capabilities. • The adequacy of the risk communication plan is validated regularly.
	Information dispose	Risk function		Ensure that information is disposed of in a timely, secure and appropriate manner.

Figure 63—Risk Communication Plan *(cont.)*				
Quality Subdimension and Goals		Description—The extent to which information is...	Relevance	Goal—The risk communication plan...
Goals — **Intrinsic**	**Accuracy**	correct and reliable	High	should accurately define type, frequency and recipient of risk communications
	Objectivity	unbiased, unprejudiced and impartial	High	information is based on the enterprise's risk culture
	Believability	regarded as true and credible	Medium	should be realistic
	Reputation	regarded as coming from a true and credible source	High	is devised based on input from both the risk function and business process owners, making the devised risk communication plan appropriate
Contextual and Representational	**Relevancy**	applicable and helpful for the task at hand	High	shall be aligned to the needs of the recipients
	Completeness	not missing and is of sufficient depth and breadth for the task at hand	High	shall cover the end-to-end enterprise structure as well external stakeholders
	Currency	sufficiently up to date for the task at hand	High	shall not be older than one year
	Amount of information	appropriate for the task at hand	High	should contain communication frequencies, types and recipients
	Concise representation	compactly represented	Medium	depends on target audience
	Consistent representation	presented in the same format	High	information shall always be presented according to the predefined formats/templates
	Interpretability	in appropriate languages, symbols, and units, and the definitions are clear	High	should be understandable for target audience
	Understandability	easily comprehended	High	should be comprehensible for target audience
	Manipulation	easy to manipulate and apply to different tasks	Medium	is tailored to cover the needs of recipients performing different tasks
Security	**Availability**	available when required, or easily and quickly retrievable	High	shall be available at required frequency to key stakeholder. Information to risk function should be available 24/7.
	Restricted access	restricted appropriately to authorised parties	High	indicates that the access to this information item is determined by the risk function, and is restricted as follows: • Write access: Risk function • Read access: All other stakeholders

	Figure 63—Risk Communication Plan *(cont.)*		
	Attribute	**Description**	**Value**
Good Practice	**Physical**	Information carrier/media	The information carrier for the risk communication plan can be an electronic and/or printed document.
	Empiric	Information access channel	The risk communication plan is accessible through the document management system, through the ERM portal and/or printed in a specific location.
	Syntactic	Code/language	The risk communication plan contains the following subparts: • Frequency/timing of communication • Type of communication • Target audience and distribution
	Semantic	Information type	Structured document providing a clear overview of the required communication.
		Information currency	Information should be current for organisational use.
		Information level	The risk communication plan will detail the required information to be communicated, depending on the target audience.
	Pragmatic	Retention period	According to data retention policy (and taking into account local legal retention requirements)
		Information status	The risk communication plan produced would be current, until a new plan is produced. Once it is replaced, it becomes the older version (historical).
		Novelty	It is updated only when required. Though the plan will be current.
		Contingency	Enterprise risk portal should provide predefined availability.
	Social	Context	The risk communication plan should be used in the context of the IT risk communication plan.
Link to Other Enablers			
Processes		The risk communication plan is an <u>output</u> from the process activities: • EDM03.03 Monitor risk management. • EDM05.01 Evaluate stakeholder reporting requirements. • EDM05.02 Direct stakeholder communication and reporting. • EDM05.03 Monitor stakeholder communication. The risk communication plan is an <u>input</u> for the process activities: • APO12.01 Collect data. • APO12.02 Analyse risk. • APO12.04 Articulate risk. • APO12.06 Respond to risk.	
Organisational Structures		The following roles are accountable and responsible for (partially) producing the information: • Accountable: ERM committee • Responsible: Risk function • Consulted: Business process owners/CIO, board, executive management, compliance, audit committee, internal audit (see also stakeholders)	
Infrastructure, Applications and Services		The risk communication plan is produced by normal office automation software.	
People, Skills and Competencies		The risk universe, appetite and tolerances require the following competencies: • Analytical capability • Interpersonal capabilities • Lateral thinking (thinking outside of the box)	
Culture, Ethics and Behaviour		The risk communication plan requires the following behaviours: • Risk-enabled/embedded culture • Positive behaviour towards raising issues • Acceptance of risk • Acceptance of risk ownership by business	
Principles, Policies and Frameworks		Related principles: • Promote fair and open communication • Consistent approach	

Figure 64—Risk Report			
A risk report includes information on current risk management capabilities and actual status and trends with regard to risk. This report will be based on the risk profile and will be tailored to the requirements of recipients.			

	Life Cycle Stage	Internal Stakeholder	External Stakeholder	Description/Stake
Life Cycle and Stakeholders	Information planning	Risk function		Risk management capabilities, actual status and trends are communicated timely and accurately to the right people, based on their needs and requirements.
	Information design	Risk function		Risk management capabilities, actual status and trends are communicated timely and accurately to the right people, based on their needs and requirements.
	Information build/acquire	Risk function, business process owners/CIO		Ensure that the risk report includes the latest information on risk management capabilities, latest status and trends with regard to risk of the entire organisation.
	Information use/operate: store, share, use	Board, executive management, CIO, risk function, business process owners/CIO, compliance, internal audit	External audit, regulator	Effective utilisation and availability of information item by all involved stakeholders.
	Information monitor	Board, risk committee, audit committee, risk function, business process owners/CIO	External audit	Verifies that information remains actual and alerts on changes.
	Information dispose	Risk function		Ensures information is disposed of in a timely, secure and appropriate manner.

	Quality Subdimension and Goals		Description—The extent to which information is...	Relevance	Goal—The risk report...
Goals	**Intrinsic**	Accuracy	correct and reliable	High	should accurately define the risk management capabilities and actual status and trends with regard to risk, in such a way that it does not arouse confusion
		Objectivity	unbiased, unprejudiced and impartial	High	information is based on the enterprise's risk culture and confirmed by observations.
		Believability	regarded as true and credible	Medium	should be realistic and accurate
		Reputation	regarded as coming from a true and credible source	High	source information is collected from competent and recognised sources.

Figure 64—Risk Report (cont.)				
Quality Subdimension and Goals		**Description—The extent to which information is…**	**Relevance**	**Goal—The risk report…**
Goals — Contextual and Representational	Relevancy	applicable and helpful for the task at hand	High	is tailored according to the requirements of the target audience
	Completeness	not missing and is of sufficient depth and breadth for the task at hand	High	typically covers the end-to-end enterprise structure as well external stakeholders
	Currency	sufficiently up to date for the task at hand	High	is typically not older than one year because it should be kept up to date
	Amount of information	appropriate for the task at hand	High	should contain an appropriate amount of information, based on the requirements of the recipients
	Concise representation	compactly represented	Medium	depends on the target audience
	Consistent representation	presented in the same format	High	shall always be presented according to the predefined formats/templates
	Interpretability	in appropriate languages, symbols, and units, and the definitions are clear	High	should be understandable for target audience
	Understandability	easily comprehended	High	should be comprehensible for target audience
	Manipulation	easy to manipulate and apply to different tasks	Medium	information can be manipulated to present in different types of risk reports
Security	Availability	available when required, or easily and quickly retrievable	High	shall be available at required frequency to the stakeholders
	Restricted access	restricted appropriately to authorised parties	High	access is determined by the risk function, and is restricted as follows: • Write access: Risk function • Read access: All other stakeholders

	Attribute	**Description**	**Value**
Good Practice	Physical	Information carrier/media	The information carrier for the risk report can be an electronic and/or printed document.
	Empiric	Information access channel	The risk report is accessible through the ERM portal and/or printed in a specific location.
	Syntactic	Code/language	The risk report contains the following subparts: • Status of risk items • Action items (risk response) • Risk capabilities
	Semantic	Information type	Information (risk) reports
		Information currency	Information should be current for organisational use.
		Information level	Depends on target audience; for board and risk committee function level a status of risk and a risk capability presentation. For process owners a detailed risk analysis report.
	Pragmatic	Retention period	According to data retention policy (and taking into account local legal retention requirements).
		Information status	The information produced would be current. Once replaced it becomes historical.
		Novelty	It is current as the risk status is updated in time.
		Contingency	Enterprise risk portal should provide as and when required.
	Social	Context	This information item should be used in context of IT risk communication plan.

Figure 64—Risk Report *(cont.)*	
Link to Other Enablers	
Processes	The risk report is an <u>output</u> from the process activities: • APO12.01 Collect data. • APO12.02 Analyse risk. • APO12.04 Articulate risk. • APO12.06 Respond to risk. The risk report is an <u>input</u> for the process activities: • EDM03.03 Monitor risk management. • EDM05.01 Evaluate stakeholder reporting requirements. • EDM05.02 Direct stakeholder communication and reporting. • EDM05.03 Monitor stakeholder communication.
Organisational Structures	The following roles are accountable and responsible for (partially) producing the information • Accountable: Risk function • Responsible: Risk function, business process owner/CIO • Consulted/Informed: Board, ERM committee, executive management, compliance, audit committee, internal audit (see also stakeholders)
Infrastructure, Applications and Services	The risk report is produced by risk management portal (GRC tool).
People, Skills and Competencies	The generation and use of the risk report requires understanding of risk management principles and skills.
Culture, Ethics and Behaviour	Risk reports require the following behaviours: • Risk-enabled/embedded culture
Principles, Policies and Frameworks	Related principles: • Align with ERM • Promote fair and open communication • Function as part of daily activities • Consistent approach

Figure 65—Risk Awareness Programme				
A risk awareness programme is a clearly and formally defined plan, with a structured approach, containing a set of related activities and procedures with the objective of realizing and maintaining a risk-aware culture.				
	Life Cycle Stage	**Internal Stakeholder**	**External Stakeholder**	**Description/Stake**
Life Cycle and Stakeholders	**Information planning**	ERM committee		• Ensures that all stakeholders understand risk in general and expectation. • Creates a risk-enabled/embedded culture. • Ensures that all risk decisions are informed.
	Information design	Risk function		Designs an effective programme for education and awareness on risk-related issues amongst staff.
	Information build/acquire	Risk function		Ensures that the risk awareness programme includes activities and procedures reaching all relevant stakeholders.
	Information use/operate: store, share, use	All		Ensures that risk-related issues are understood.
	Information monitor	Internal audit, risk function		Monitors that the plan remains adequate to continue to create appropriate risk awareness at all levels of the enterprise.
	Information dispose	Risk function		Ensures information is disposed of in a timely, secure and appropriate manner.

	Quality Subdimension and Goals		Description—The extent to which information is...	Relevance	Goal—The risk awareness programme...
Goals	Intrinsic	**Accuracy**	correct and reliable	High	is based on standard risk-related education
		Objectivity	unbiased, unprejudiced and impartial	High	is based on the organisation's risk culture and appetite
		Believability	regarded as true and credible	High	is customised to the audience
		Reputation	is regarded as coming from a true and credible source	High	is prepared based on the approved enterprise risk policy
	Contextual and Representational	**Relevancy**	applicable and helpful for the task at hand	High	is organised according to target audience
		Completeness	not missing and is of sufficient depth and breadth for the task at hand	High	typically covers the end-to-end enterprise structure as well external stakeholders
		Currency	sufficiently up to date for the task at hand	Medium	should be verified for appropriateness every six months
		Amount of information	appropriate for the task at hand	Medium	includes basic information about risk management and expectation from key stakeholders
		Concise representation	compactly represented	Medium	depends on target audience
		Consistent representation	presented in the same format	Medium	shall always be presented according to the predefined formats/templates and target audience
		Interpretability	in appropriate languages, symbols, and units, and the definitions are clear	Medium	should be understandable for target audience
		Understandability	easily comprehended	Medium	should be comprehensible for target audience
		Manipulation	Easy to manipulate and apply to different tasks	Medium	activities and procedures should be adaptable, depending on the target audience
	Security	**Availability**	Available when required, or easily and quickly retrievable	Medium	shall be available at required frequency (risk workshops) to key stakeholders
		Restricted access	Access is restricted appropriately to authorised parties	Low	access is determined by the risk function, and is restricted as follows: • Write access: Risk function • Read access: All other stakeholders

	Attribute	**Description**	**Value**
Good Practice	**Physical**	Information carrier/media	The information carrier for the risk awareness programme can be an electronic and/or printed document.
	Empiric	Information access channel	The risk awareness programme is accessible through document management system/enterprise risk portals and through workshops.
	Syntactic	Code/language	The risk awareness programme contains risk-related education, frequency and audience.
	Semantic	Information type	Risk awareness/education.
		Information currency	The risk awareness programme should be current for organisational use.
		Information level	A programme with frequency, content and target audience
	Pragmatic	Retention period	According to data retention policy (and taking into account local legal retention requirements)
		Information status	The information produced would be current. Once replaced it becomes historical.
		Novelty	It is current until it is updated.
		Contingency	Enterprise risk portal should provide as and when required.
	Social	Context	The risk awareness programme should be used in context of a risk education requirement.

Figure 65—Risk Awareness Programme *(cont.)*

Figure 65—Risk Awareness Programme *(cont.)*	
Link to Other Enablers	
Processes	The risk awareness programme is an <u>output</u> from the process activity: • EDM03.02 Direct risk management. The risk awareness programme is an <u>input</u> for the process activity: • APO012 Manage risk.
Organisational Structures	The following roles are accountable and responsible for (partially) producing the information: • Accountable: Risk function • Responsible: Risk function • Consulted: Board, ERM committee, executive management, compliance, audit committee, internal audit • Informed: All (see also stakeholders)
Infrastructure, Applications and Services	The risk awareness programme is produced by a document management tool and/or an enterprise risk portal.
People, Skills and Competencies	The generation and use of the risk awareness programme requires understanding of risk management principles and skills.
Culture, Ethics and Behaviour	The risk awareness programme requires setting a basic risk culture and recognition of risk exposure of the organisation.
Principles, Policies and Frameworks	Related principles: • Promote fair and open communication • Function as part of daily activities

Figure 66—Risk Map			
A common, very easy and intuitive technique to present risk is the risk map, where risk is plotted on a two-dimensional diagram, with frequency and impact being the two dimensions. The risk map representation is powerful and provides an immediate and complete view on risk and apparent areas for action. Furthermore, a risk map allows defining colour zones indicating appetite bands of significance in graphical mode. A practical consideration would be to focus on the 10 or 20 most important risk items, or split them up into categories, as the map would otherwise not be readable.			
Life Cycle Stage	**Internal Stakeholder**	**External Stakeholder**	**Description/Stake**
Information planning	Risk function		Risk is communicated in a timely manner.
Information design	Risk function		• Design risk maps according to the ERM committee/executive management requirements. • Ensures that risk is presented in an understandable manner.
Information build/acquire	Risk function, business process owners/CIO		Gather the required information from the key stakeholders to populate the risk map.
Information use/operate: store, share, use	Board, executive management, CIO, risk function, business process owners/CIO, compliance, internal audit	External audit, regulator	Effective utilisation of the risk maps to support action taking and decision making.
Information monitor	Board, risk committee, audit committee, risk function, business process owners/CIO	External audit	• Monitor whether the information on the risk map is still based on the most up to date information. • Ensure that the risk map is still relevant for the enterprise.
Information dispose	Risk function		Ensure that information is disposed of in a timely, secure and appropriate manner.

(Left vertical label: Life Cycle and Stakeholders)

	Quality Subdimension and Goals		Description—The extent to which information is…	Relevance	Goal—The risk map…
Goals	**Intrinsic**	**Accuracy**	correct and reliable	High	should accurately define risk status of all risk items
		Objectivity	unbiased, unprejudiced and impartial	High	information is collected/confirmed by multiple sources in the enterprise and external to the enterprise, if applicable. The information can be stored in a GRC tool.
		Believability	regarded as true and credible	Medium	view should be realistic, apparent areas for action considered, taking into account the current state and environment of the enterprise
		Reputation	regarded as coming from a true and credible source	High	source information is collected from competent and recognised sources
	Contextual and Representational	**Relevancy**	applicable and helpful for the task at hand	High	shall be organised according risk categories to ensure adequate comparison possibilities
		Completeness	not missing and is of sufficient depth and breadth for the task at hand	High	shall cover the end-to-end enterprise risk spectrum
		Currency	sufficiently up to date for the task at hand	High	shall present current status
		Amount of information	appropriate for the task at hand	High	should contain very high-level information. Volume of source information should be appropriate to support the indication in the risk map.
		Concise representation	compactly represented	Medium	information shall always be presented according to the predefined formats/templates
		Consistent representation	presented in the same format	High	information shall always be presented according to the predefined formats/templates
		Interpretability	in appropriate languages, symbols, and units, and the definitions are clear	High	should be understandable for target audience
		Understandability	easily comprehended	High	should be comprehensible for target audience
		Manipulation	easy to manipulate and apply to different tasks	Medium	information can be manipulated to present it into different graphic methods
	Security	**Availability**	available when required, or easily and quickly retrievable	High	shall be available 24/7
		Restricted access	restricted appropriately to authorised parties	High	access is determined by the risk function, and is restricted as follows: • Write access: Risk function • Read access: All other stakeholders

	Attribute	**Description**	**Value**
Good Practice	**Physical**	Information carrier/media	The information carrier for the risk map can be an electronic and/or printed document.
	Empiric	Information access channel	The risk map is accessible through the ERM portal.
	Syntactic	Code/language	The risk map contains the following subparts: • Status of risk items (indicated on the risk map) • Areas for action and, optionally, a risk response
	Semantic	Information type	Risk status in graphical format
		Information currency	Information should be current for organisational use.
		Information level	Graphical representation of risk status and area for action
	Pragmatic	Retention period	According to data retention policy (and taking into account local legal retention requirements)
		Information status	The risk map produced would be current. Once replaced it becomes historical.
		Novelty	It is current as the risk map is updated in time.
		Contingency	Enterprise risk portal should provide as and when required.
	Social	Context	The risk map should be used in the context of the IT risk communication plan.

Figure 66—Risk Map *(cont.)*

Figure 66—Risk Map (cont.)	
Link to Other Enablers	
Processes	The risk map is an <u>output</u> from the process activities: • APO12.01 Collect data. • APO12.02 Analyse risk. • APO12.04 Articulate risk. • APO12.06 Respond to risk. The risk map is an <u>input</u> for the process activities: • EDM05.01 Evaluate stakeholder reporting requirements. • EDM05.02 Direct stakeholder communication and reporting. • EDM05.03 Monitor stakeholder communication. • EDM03.03 Monitor risk management.
Organisational Structures	The following roles are accountable and responsible for (partially) producing the information • Accountable: Risk function • Responsible: Risk function, business process owners/CIO • Consulted/Informed: Board, ERM committee, executive management, compliance, audit committee, internal audit (see also stakeholders)
Infrastructure, Applications and Services	The risk map is produced by the risk management portal (GRC tool).
People, Skills and Competencies	The generation and use of the risk map requires understanding of risk management principles and skills.
Culture, Ethics and Behaviour	Use of the risk map requires a risk-enabled/embedded culture.
Principles, Policies and Frameworks	Related principles: • Connect to enterprise objectives • Align with ERM • Balance cost/benefit of IT risk

Figure 67—Risk Universe, Appetite and Tolerance				
• **Universe**—The full amount of risk, including the unknowns, which could have an impact, either positively or negatively, on the ability of an enterprise to achieve its long-term mission (or vision). • **Appetite**—The broad-based amount of risk, in different aspects of an enterprise, that it is willing to accept in pursuit of its mission. • **Tolerance**—The acceptable level of variation that management is willing to allow for any particular risk as it pursues objectives.				
	Life Cycle Stage	**Internal Stakeholder**	**External Stakeholder**	**Description/Stake**
Life Cycle and Stakeholders	**Information planning**	Board, CEO and ERM committee		Stakeholders define the universe, appetite and tolerance.
	Information design	Risk function		Designs the appropriate and clear cut description of the universe and appetite and defines the required information to be able to set the tolerances.
	Information build/acquire	Risk function		Gathers the (remaining) required information to set the universe, appetite and tolerances.
	Information use/operate: store, share, use	Board, executive management, risk function, business process owners/CIO, compliance, internal audit	External audit, regulator	Effective utilisation and availability of the risk universe, appetite and tolerances.
	Information monitor	Board, risk committee, audit committee, risk function, business process owners/CIO	External audit	Ensures risk acceptance remains within acceptable limits and adapts the risk universe, appetite and tolerances if required.
	Information dispose	Risk function		Ensures information is disposed of in a timely, secure and appropriate manner.

	Quality Subdimension and Goals		Description—The extent to which information is…	Relevance	Goal—The risk universe, appetite and tolerance…
Goals	**Intrinsic**	**Accuracy**	correct and reliable	High	should accurately articulate thresholds
		Objectivity	unbiased, unprejudiced and impartial	High	information is collected/confirmed by multiple sources in the enterprise and external to the enterprise if applicable
		Believability	regarded as true and credible	Medium	should be in line with the enterprise's risk culture
		Reputation	regarded as coming from a true and credible source	High	source information is collected from a workshop with all required stakeholders
	Contextual and Representational	**Relevancy**	applicable and helpful for the task at hand	Medium	are maintained in preapproved templates.
		Completeness	not missing and is of sufficient depth and breadth for the task at hand	High	descriptions shall consider all required information to accurately describe the boundaries
		Currency	sufficiently up to date for the task at hand	High	shall be present current status
		Amount of information	appropriate for the task at hand	High	should provide relevant information for informed decisions
		Concise representation	compactly represented	Medium	shall always be presented according to the predefined formats/templates
		Consistent representation	presented in the same format	High	shall always be presented according to the predefined formats/templates
		Interpretability	in appropriate languages, symbols, and units, and the definitions are clear	High	should be understandable for target audience
		Understandability	easily comprehended	Medium	should be comprehensible for target audience
		Manipulation	easy to manipulate and apply to different tasks	Medium	should be described in such a way that is clear what it means specifically for all departments, functions, jobs, etc., in the enterprise.
	Security	**Availability**	available when required, or easily and quickly retrievable	High	shall be available 24/7
		Restricted access	restricted appropriately to authorised parties	High	access is determined by the risk function, and is restricted as follows: • Write access: Risk function • Read access: All other stakeholders

	Attribute	Description	Value
Good Practice	**Physical**	Information carrier/media	The information carrier for the risk universe, appetite and tolerance can be an electronic and/or printed document.
	Empiric	Information access channel	The risk universe, appetite and tolerance item is accessible through the ERM portal.
	Syntactic	Code/language	The risk universe, appetite and tolerance contain historical financial information, tolerance/acceptance level and a number of transactions/activities.
	Semantic	Information type	Standard template
		Information currency	Information should be current for organisational use.
		Information level	Detailed descriptions of the risk universe, appetite and tolerance, applicable for the scope of the organisation
	Pragmatic	Retention period	According to data retention policy (and taking into account local legal retention requirements)
		Information status	The risk universe, appetite and tolerance produced would be current. Once replaced it becomes historical.
		Novelty	It is current as the risk status is updated in time.
		Contingency	Enterprise risk portal should provide as and when required.
	Social	Context	The risk universe, appetite and tolerance should be used in the context of the risk policy.

Figure 67—Risk Universe, Appetite and Tolerance (cont.)

Figure 67—Risk Universe, Appetite and Tolerance *(cont.)*	
Link to Other Enablers	
Processes	The risk universe, appetite and tolerances are an <u>output</u> from the process activities: • EDM03.01 Evaluate risk management. • EDM03.03 Monitor risk management. The risk universe, appetite and tolerances are an <u>input</u> for the process activities • APO12.03 Maintain a risk profile. • APO12.04 Articulate risk.
Organisational Structures	The following roles are accountable and responsible for (partially) producing the information • Accountable: Risk function • Responsible: Risk function • Consulted/Informed: Board, ERM committee, audit committee, business process owners/CIO, internal audit (see also stakeholders)
Infrastructure, Applications and Services	The risk universe, appetite and tolerances are made available on the risk management portal (GRC tool).
People, Skills and Competencies	The risk universe, appetite and tolerances require the following competencies: • Analytical capability • Interpersonal capabilities • Lateral thinking (thinking outside of the box)
Culture, Ethics and Behaviour	The risk universe, appetite and tolerances require the following behaviours: • Positive behaviour towards raising issues • Risk acceptance as a valid option • Ownership of risk is accepted by business
Principles, Policies and Frameworks	Related principles: • Connect to enterprise objectives • Align with ERM • Balance cost/benefit of IT risk • Establish tone at the top and accountability

About Risk Appetite, Risk Capacity and Risk Tolerance

In the discussion about risk appetite, the term 'risk capacity' is sometimes used as well. This is usually defined as the objective amount of loss that an enterprise can tolerate without risking its continued existence. As such, it differs from risk appetite, which is more a board/management decision on how much risk is desirable, as shown in **figure 68**.

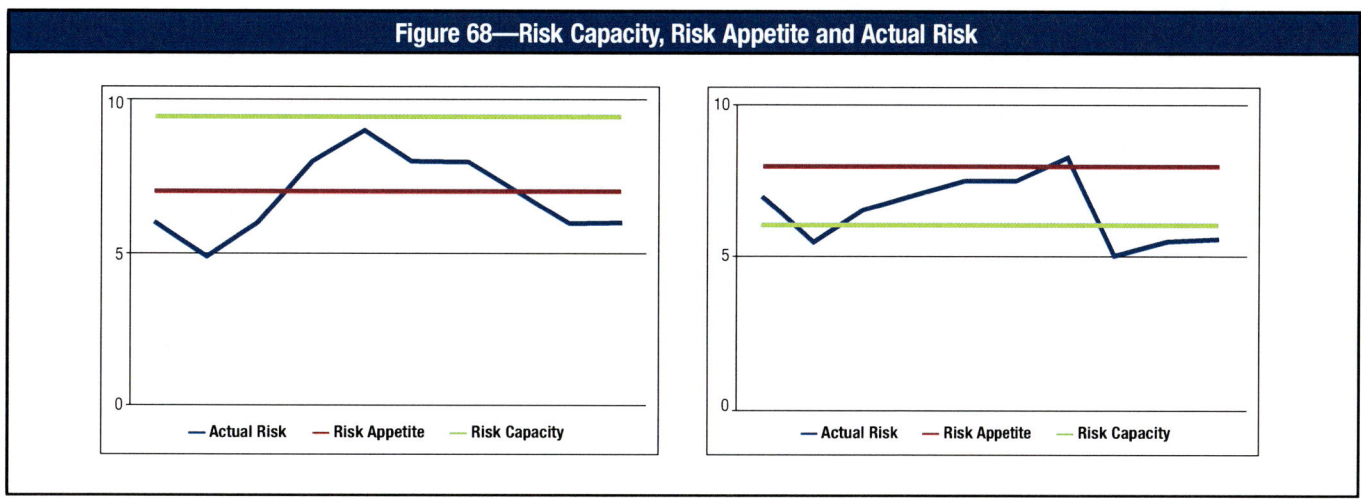

Figure 68—Risk Capacity, Risk Appetite and Actual Risk

In **figure 68** the:
• The left diagram shows a relatively sustainable situation where risk appetite is lower than risk capacity, and where actual risk exceeds risk appetite in a number of situations, but always remains below the risk capacity
• The right diagram shows a rather unsustainable situation, where risk appetite is defined by management at a level beyond risk capacity; this means that management is prepared to accept risk well over the objective capacity to absorb loss. As a result, actual risk routinely exceeds risk capacity even when staying almost always below the risk appetite level. This usually represents an unsustainable situation.

Defining Risk Capacity and Risk Appetite

Risk capacity and risk appetite are defined by board and executive management at the enterprise level (EDM03). There are several benefits associated with this approach:

• Supporting and providing evidence of the risk-based decision-making processes, because all risk decisions are based on where the risk resides on the risk map; hence, all risk response actions can be tracked back and justified
• Supporting the understanding of how each component of the enterprise contributes to the overall risk profile
• Showing how different resource allocation strategies can add to or lessen the burden of risk, by simulating different risk response options
• Supporting the prioritisation and approval process of risk response actions through risk budgets; a risk budget allows enterprises to trade off types of risk (time to market vs. reliability) and risk acceptance vs. investment to reduce. For example, it is helpful to understand how to use a risk budget to spend on mitigating 'known' risk items (e.g., operational stability and availability of an ordering infrastructure) to allow acceptance of unknown risk (e.g., new product take-on rates)
• Identifying specific areas where a risk response should be made

Risk appetite is translated into a number of standards and policies, in order to contain the risk level within the boundaries set by the risk appetite, for example:

• Management of a financial service firm has determined that the main processing platform and applications cannot be unavailable for any period longer than two hours, and that the system should be able to process yearly transaction growth of 15 percent without performance impact. IT management then needs to translate this into specific available and redundancy requirements for the servers and other infrastructure the applications are running on. In turn this leads to:
  –Detailed technical capacity requirements and forecasts requirements
  –Specific IT procedures for performance monitoring and capacity planning
• Management has determined that new business initiatives' time to market is crucial and that IT applications supporting these initiatives cannot be delivered with delays exceeding one month—no exceptions allowed. IT management will then have to translate this into resource requirements and development process requirements for all development initiatives.

Similar to the risk universe description and enterprise risk assessments, the risk appetite and the boundaries between different bands of significance need to be regularly adjusted or confirmed.

Risk Tolerance

Risk tolerance levels are tolerable deviations from the level set by the risk appetite definitions, for example:

• Standards require projects to be completed within the estimated budgets and time, but an overrun of 10 percent budget or 20 percent time are tolerated.
• Service levels for system uptime require 99.5 percent availability on a monthly basis; however, isolated cases of 99.4 percent will be tolerated.
• The enterprise is very security risk averse and does not want to accept any external intrusions; however single isolated intrusions with limited damage can be tolerated.
• A user profile approval procedure exists, but in some instances if the procedure is not fully complied with it can be tolerated.

In the previous examples, risk tolerance is defined using IT process metrics or adherence to defined IT procedures and policies, which in turn are a translation of the IT goals that need to be achieved.

Risk appetite and risk tolerance go hand in hand. Risk tolerance is defined at the enterprise level and is reflected in policies set by the executives; at lower (tactical) levels of the enterprise, or in some entities of the enterprise, exceptions can be tolerated (or different thresholds defined), as long as at the enterprise level the overall exposure does not exceed the set risk appetite. Any business initiative includes a risk component, so management should have the discretion to exceed risk tolerance levels and pursue some new opportunities. In enterprises where policies are cast in granite rather than 'lines in the sand,' the agility and innovation to exploit new business opportunities could be lacking. Conversely, there are situations where policies are based on specific legal, regulatory or industry requirements where it is appropriate to have no risk tolerance for failure to comply.

Risk appetite and tolerance should be defined and approved by senior management and clearly communicated to all stakeholders. A process should be in place to review and approve any exceptions to such standards. Risk appetite is the more static translation on how much risk generally is acceptable; risk tolerance allows individual and justified exceptions.

Risk appetite and tolerance change over time. New technology, new organisational structures, new market conditions, new business strategy and many other factors require the enterprise to re-assess its risk portfolio at regular intervals, and also require the enterprise to re-confirm its risk appetite at regular intervals, triggering risk policy reviews. In this respect, an enterprise needs also to understand that the better risk management it has in place, the more risk that can be taken in pursuit of return.

The cost of mitigation options can affect risk tolerance; indeed, there may be circumstances where the cost/business impact of risk mitigation options exceeds an enterprise's capabilities/resources, thus forcing higher tolerance for one or more risk conditions. For example, if a regulation says that 'sensitive data at rest must be encrypted', yet there is no feasible encryption solution or the cost of implementing a solution is grossly impactful, then the enterprise may choose to accept the risk associated with regulatory non-compliance.

	Figure 69—What Is a Key Risk Indicator?			
	Risk indicators are metrics capable of showing that the enterprise is, or has a high probability of being, subject to a risk that exceeds the defined risk appetite. They are specific to each enterprise, and their selection depends on a number of parameters in the internal and external environment, such as the size and complexity of the enterprise, whether it is operating in a highly regulated market, and its strategy focus. A key risk indicator (KRI) is differentiated as being highly relevant and possessing a high probability of predicting or indicating important risk.			
	Life Cycle Stage	**Internal Stakeholder**	**External Stakeholder**	**Description/Stake**
Life Cycle and Stakeholders	**Information planning**	ERM committee		Elaborates on a high-level description of the risk indicators needed to measure the enterprise's risk appetite.
	Information design	Risk function,		Elaborates the risk indicators in more detail to come to a detailed description.
	Information build/acquire	Risk manager, risk analyst		Build the risk indicators in full detail (identify required information) and make sure they are ready to use (set up information feeding streams).
	Information use/operate: store, share, use	Board, executive management, risk function, business process owners/CIO, compliance, internal audit	External audit, regulator	Effective use of the risk indicators
	Information monitor	Board, ERM committee, audit committee, risk function, business process owners/CIO	External audit	Monitors the risk indicators to make sure they stay in between the defined boundaries, according to the risk appetite of the enterprise.
	Information dispose	Risk function		Evaluates the relevance of a risk indicator on a timely basis.

			Figure 69—Key Risk Indicators *(cont.)*		
		Quality Subdimension and Goals	**Description—The extent to which information is…**	**Relevance**	**Goal—Risk indicators…**
Goals	**Intrinsic**	**Accuracy**	correct and reliable	High	need to be accurate and should accurately articulate thresholds
		Objectivity	unbiased, unprejudiced and impartial	High	should be based on objective information—measurable components
		Believability	regarded as true and credible	High	should indicate that people are convinced of the correct definition and contribution to the insight in the risk exposure of the enterprise
		Reputation	regarded as coming from a true and credible source	Medium	information source should be clearly mentioned and visible for the users and should come from a trusted enterprise source
	Contextual and Representational	**Relevancy**	applicable and helpful for the task at hand	High	should indicate that people are convinced of the contribution to the insight of the exposure of the enterprise
		Completeness	not missing and is of sufficient depth and breadth for the task at hand	High	should contain all the key elements that have an influence on the risk that the indicator is quantifying
		Currency	sufficiently up to date for the task at hand	High	individual source elements need to be as up to date as possible to have an up-to-date indication of the exposure
		Amount of information	appropriate for the task at hand	Low	depend on the number of individual elements
		Concise representation	compactly represented	Low	are metrics and thus compactly represented by a figure. On the other hand, the explanation on the indicator and on the possible consequences can be comprehensive.
		Consistent representation	presented in the same format	Medium	are represented in a standard format, although specific alterations can be made for some indicators
		Interpretability	in appropriate languages, symbols, and units, and the definitions are clear	High	composition and goal/meaning should be clear to the target audience
		Understandability	easily comprehended	High	composition and goal/meaning should be clear to the target audience
		Manipulation	easy to manipulate and apply to different tasks	Low	can only be presented in a limited number of ways, and changing the presentation should not impact how the indicators assess the exposure
	Security	**Availability**	available when required, or easily and quickly retrievable	Medium	shall be available at all times, as some of them might be linked to fast fluctuating business indicators (e.g., stock exchange)
		Restricted access	restricted appropriately to authorised parties	Low	access: • Everybody can consult the general indicators. Access to more specific indicators in restricted. • A selective number of people across the enterprise will gather the needed information and populate the indicators. • The ERM committee and the risk function will have modification rights to the composition of the indicators.

Figure 69—Key Risk Indicators *(cont.)*			
	Attribute	**Description**	**Value**
Good Practice	**Physical**	Information carrier/media	The information carrier for the risk indicators can be an electronic and/or printed document
	Empiric	Information access channel	The risk indicators are accessible through the ERM portal (with restricted access implemented).
	Syntactic	Code/language	The risk indicators contain metrics and an explanation on the goal/meaning.
	Semantic	Information type	Fixed composition of a risk indicator
		Information currency	The risk indicators should be current for organisational use.
		Information level	The risk indicators collect data over the entire enterprise to show the exposure of a certain process, product, or department.
	Pragmatic	Retention period	As long as the indicator still gives a correct degree of exposure—meaning, if all the elements are still present that are key influencers in the degree of exposure being measured—it continues to be retained.
		Information status	The information produced would be current.
		Novelty	It is current as it uses the latest status of all the elements in the indicator.
		Contingency	This information relies on the following information being available and understood by the user: • Risk appetite of the enterprise • The composition of the risk indicator • The indication it provides on the risk appetite of the enterprise
	Social	Context	This information is useful in verifying if a certain exposure is exceeding the defined risk appetite.
Link to Other Enablers			
Processes		The risk indicators are an <u>output</u> from the process activities: • EDM03.01 Evaluate risk management. • EDM03.03 Monitozr risk management. The risk indicators are an <u>input</u> for the process activities: • APO12.03 Maintain a risk profile. • APO12.04 Articulate risk.	
Organisational Structures		The following roles are accountable and responsible for (partially) producing the information: • Accountable: Risk function • Responsible: Risk manager, risk analyst • Consulted/Informed: Board, ERM committee, audit committee, business process owners/CIO, internal audit (see also stakeholders)	
Infrastructure, Applications and Services		The risk indicators are produced via the risk management portal (GRC tool).	
People, Skills and Competencies		The risk indicators require the following competencies: • Analytical capability • Lateral thinking (thinking outside of the box) • Technical understanding • Risk expertise • Organisational and business awareness.	
Culture, Ethics and Behaviour		The risk indicators require the following behaviours: • KRIs are used as an early warning • Risk indicators or events that fall outside of tolerance are acted on • Risk trends are reported to management • Ownership of risk is accepted by the business • Recognition of exposure • Transparent behaviours • Showing effort to understand what risk is for each stakeholder and how it impacts their objectives	
Principles, Policies and Frameworks		Related principles: • Establish tone at the top and accountability • Promote fair and open communication	

Key Risk Indicators

Any metric that can be used to describe and track a risk is an indicator of that risk. Risk indicators are specific to each enterprise, i.e., many risk indicators are available and some are more appropriate to specific enterprises. Their selection depends on a number of parameters in the internal and external environment, such as the size and complexity of the enterprise, whether it is operating in a highly regulated market, and its strategy focus. Identifying risk indicators should take into account the following aspects (amongst others):
• Consider the different stakeholders in the enterprise. Risk indicators should not focus solely on either the more operational or the strategic side of risk. Risk indicators can and should be identified for all stakeholders. Involving the right stakeholders in the selection of risk indicators will also ensure greater buy-in and ownership.
• Make a balanced selection of risk indicators, covering lag indicators (indicating risk after events have occurred), lead indicators (indicating what capabilities are in place to prevent events from occurring) and trends (analysing indicators over time or correlating indicators to gain insights). Ensure that the selected indicators drill down to the root cause of the events (indicative of root cause and not just symptoms).

An enterprise may develop an extensive set of metrics to serve as risk indicators; however, it is neither possible nor feasible to maintain that full set of metrics as KRIs. A KRI is differentiated as being highly relevant and possessing a high probability of predicting or indicating important risk. Criteria to select KRIs include:
• **Impact**—Indicators for risk items with high business impact are more likely to be KRIs.
• **Effort** (to implement, measure and report)—For different indicators that are equivalent in sensitivity, the one that is easier to measure is preferred.
• **Reliability**—The indicator must possess a high correlation with the risk and be a good predictor or outcome measure.
• **Sensitivity**—The indicator must be representative for risk and capable of accurately indicating variances in the risk.

To illustrate the difference between reliability and sensitivity in the previous list, an example of a smoke detector can be used. Reliability means that the smoke detector will sound an alarm every time that there is smoke, and sensitivity means that the smoke detector will sound when a certain threshold of smoke density is reached.

The complete set of KRIs should also balance both indicators for risk and root causes as well as business impact. The selection of the right set of KRIs will have the following benefits to the enterprise:
• Provide an early warning (forward-looking) signal that a high risk is emerging to enable management to take proactive action (before the risk actually becomes a loss)
• Provide a backward-looking view on risk events that have occurred, enabling risk responses and management to be improved
• Enable the documentation and analysis of trends
• Provide an indication of the enterprise risk appetite through metric setting (i.e., KRI thresholds)
• Increase the likelihood of achieving the enterprise strategic objectives
• Assist in continually optimising the risk governance and management environment

Some of the common challenges encountered when successfully implementing KRIs include:
• KRIs are not linked to specific risk items
• KRIs are often incomplete or inaccurate in specification, i.e., too generic
• Lack of alignment between risk, KRI description and KRI metric
• Having too many KRIs
• KRIs are difficult to measure
• Difficulty to aggregate, compare and interpret KRIs in a systematic fashion at an enterprise level

Because an enterprise's internal and external environment is constantly changing, the risk environment is also highly dynamic, and the set of KRIs will need to be updated regularly. Each KRI will be related to the risk appetite and tolerance so that trigger levels can be defined. This will enable stakeholders to take appropriate action in a timely manner.

In addition to indicating risk, KRIs are particularly important during the communication of risk. They facilitate the dialogue on risk within the enterprise, based on clear and measurable facts. At the same time, KRIs can be used to improve risk awareness throughout the enterprise due to the factual nature of these indicators.

In the COBIT 5 framework, adequate governance and management are achieved through the enablers; because a generic enabler performance model is available, all components of the enabler performance model are candidates to serve as KRIs, for example:
• All metrics related to the achievement of enabler goals, e.g., process goals, organisational structure goals
• All metrics related to the application of good practice for enablers, e.g., application of process practices
• All metrics related to the life cycle management of enablers
• Combined metrics, e.g., process capability levels (according to ISO/IEC 15504-based process assessments)

In addition, the COBIT 5 framework includes also the goals cascade with different levels of metrics (business process, IT process, practice) designed to measure successful operations and outcome of IT processes in support of the business. From these metrics, a selection can be made to indicate the quality of the process practices (or for that matter any other enabler) put in place to mitigate risk.

Figure 70 contains an example of some possible KRIs for different stakeholders. Both lead and lag types of indicators are used. This table is not complete (nor is it intended to be), but it provides the reader with some suggestions and inspiration for their own set of KRIs.

The stakeholders considered here are:
• CIO—This function requires a view on IT risk for the enterprise.
• Risk function—The risk function requires a broad view on IT risk from across the business, but can be considered to have a more operational focus.
• CEO/Board—These entities require a high-level, aggregated view on risk.

Source[8]	Event Category	CIO	Risk Function	CEO/Board
EDM02, APO02, APO05, BAI01	Investments/project decision-related events	• Percent of projects on time, on budget • Number and type of deviations from technology infrastructure plan	• Percent of IT projects reviewed and signed off on by quality assurance (QA) that meet target quality goals and objectives • Percent of projects with benefit defined upfront	• Percent of IT investments exceeding or meeting the predefined business benefit • Percentage of IT expenditures that have direct traceability to the business strategy
EDM01, EDM02, APO02, APO05	Business involvement-related events	Degree of approval of business owners of the IT strategic/ tactical plans	Frequency of meetings with enterprise leadership involvement where IT's contribution to value is discussed	Frequency of CIO reporting to or attending executive board meetings at which IT's contribution to enterprise goals is discussed
APO13, DSS05	Security	Percent of users who do not comply with password standards	Number and type of suspected and actual access violations	Number of (security) incidents with business impact
APO07, DSS01, DSS02, DSS03, DSS04, BAI08, MEA03	Involuntary staff act: Destruction	• Number of service levels impacted by operational incidents • Percent of IT staff who complete annual IT training plan	• Number of incidents caused by deficient user and operational documentation and training • Number of business-critical processes relying on IT not covered by IT continuity plan	• Cost of IT non-compliance, including settlements and fines • Number of non-compliance issues reported to the board or causing public comment or embarrassment

Figure 70—Example KRIs

---

[8] The source refers to the COBIT 5 process references.

Figure 71—Emerging Risk Issues and Factors			
Emerging risk issue and factors are information on an upcoming or likely combination of control, value and threat condition that imposes a noteworthy level of future IT risk.			
Life Cycle Stage	**Internal Stakeholder**	**External Stakeholder**	**Description/Stake**
Information Planning	ERM committee, CIO, risk function	External audit, regulator	Elaborates on a high-level description of the emerging risk issues and factors considered relevant to the enterprise.
Information Design	CIO, risk function		Elaborates the emerging risk issues and factors in more detail to come to a thorough understanding.
Information build/acquire	CIO, risk function, business process owners, project management office, CISO, head architect, head development, head IT operations, head IT administration, service manager, information security manager, business continuity manager, privacy officer		Build the emerging risk issues and factors in full detail (identify required organisational information) and make sure they are ready for monitoring by stakeholders.
Information use/operate: store, share, use	Risk function, CISO, service manager, CIO, head architect, head development, head IT operations, information security manager, business process owners, project management office, compliance, audit, head development, head IT administration, business continuity manager, privacy officer, COO, business executives	External audit, regulator	Effective monitoring by all stakeholders of the emerging threats identified as relevant for the enterprise.
Information monitor	CIO, audit	External audit	Register observations by the stakeholders regarding the emerging issues and factors and update the assessment of them accordingly.
Information dispose	CIO, risk function		Evaluates the relevance of the emerging issues and factors on a timely basis.

Life Cycle and Stakeholders

Figure 71—Emerging Risk Issues and Factors *(cont.)*					
		Quality Subdimension and Goals	**Description—The extent to which information is…**	**Relevance**	**Goal—Emerging risk issues and factors…**
Goals	Intrinsic	**Accuracy**	correct and reliable	High	source information needs to be accurate (confirm through audit) and needs to be processed, analysed and presented in the risk management application according to fixed rules
		Objectivity	unbiased, unprejudiced and impartial	High	information is based on verifiable facts and substantiations, using the common risk view established throughout the enterprise
		Believability	regarded as true and credible	Medium	can be so new to the enterprise, especially in high-tech markets, that not all stakeholders will regard the information as true or credible
		Reputation	regarded as coming from a true and credible source	High	source information is collected from competent and recognised sources.
	Contextual and Representational	**Relevancy**	applicable and helpful for the task at hand	Medium	are structured according to the 'syntax' attribute, and it shall be confirmed by the CIO and CRO (APO12) that the information is relevant
		Completeness	not missing and is of sufficient depth and breadth for the task at hand	High	shall cover the full enterprise and the full risk register
		Currency	sufficiently up to date for the task at hand	High	information shall be no older than one month
		Amount of information	appropriate for the task at hand	Medium	tend to have little information available when first identified
		Concise representation	compactly represented	High	shall be concisely represented; this is obtained by collecting, processing and analysing the risk-related data for the entire enterprise and by detecting and presenting only emerging issues and factors
		Consistent representation	presented in the same format	Medium	shall always be presented according to the current template
		Interpretability	in appropriate languages, symbols, and units, and the definitions are clear	Low	will be new to many stakeholders and need to be easy comprehensible and thus use commonly used symbols and units
		Understandability	easily comprehended	High	will be new to many stakeholders and need to be easy comprehensible
		Manipulation	easy to manipulate and apply to different tasks	Medium	will, in the beginning, be considered to be applicable in many domains and thus need to be described in such a way that they apply to all the different tasks it pertains to
	Security	**Availability**	available when required, or easily and quickly retrievable	Medium	shall at all times be available to its stakeholders; an unavailability of 24 hours is acceptable in the case of incidents
		Restricted access	restricted appropriately to authorised parties	High	access is determined by the risk function and CIO, and is restricted as follows: • Write access: Risk function and CIO • Read access: All other stakeholders

	Attribute	Description	Value
Good Practice	**Physical**	Information carrier/media	The information carrier for the emerging risk issues and factors can be an electronic or printed document.
	Empiric	Information access channel	The emerging risk issues and factors are accessible through the ERM portal or printed at a defined area.
	Syntactic	Code/language	The emerging risk issues and factors contain the following subparts: • Analysed historical data • Emerging issues • Emerging factors • Risk profile relations
	Semantic	Information type	Risk report style
		Information currency	The emerging risk issues and factors analyse new or recent information and even consider potential future states.
		Information level	The emerging risk issues and factors collect data over the entire enterprise, processes and analyses the data and to detect emerging risk issues and factors.
	Pragmatic	Retention period	The emerging risk issues and factors are to be retained for as long as the data/information over which it reports risk needs to be retained (and taking into account local legal retention requirements).
		Information status	All instances produced over the last three months are operational; older ones are historical data.
		Novelty	The emerging risk issues and factors combine several other sources of data that make up a new instance; hence, it is novel data. It is updated on a weekly basis.
		Contingency	The emerging risk issues and factors rely on the following information being available and understood by the user: • Risk events over the entire enterprise • Risk appetite of the enterprise • Risk issues of the enterprise • Risk factors that apply to the enterprise • Risk taxonomy in use in the enterprise
	Social	Context	The emerging risk issues and factors are only meaningful and to be used in a context of ERM of risk.

Figure 71—Emerging Risk Issues and Factors *(cont.)*

Link to Other Enablers

Processes	The emerging risk issues and factors are an <u>output</u> from the process activity: • APO12.06 Respond to risk. The emerging risk issues and factors are an <u>input</u> for the process activities: • MEA02.07 Scope assurance initiatives. • MEA02.08 Execute assurance initiatives.
Organisational Structures	The following roles are accountable and responsible for (partially) producing the information: • Accountable: CIO • Responsible: Risk function, business process owners, project management office, CISO, head architect, head development, head IT operations, head IT administration, service manager, information security manager, business continuity manager, privacy officer
Infrastructure, Applications and Services	The emerging risk issues and factors can be produced by a risk management application and can be used by service management, monitoring, compliance and audit applications.
People, Skills and Competencies	The generation and use of this information item requires basic understanding of risk management principles and skills.
Culture, Ethics and Behaviour	Stakeholders must be educated towards determination of emerging risk issues.
Principles, Policies and Frameworks	Related principles: • Connect to enterprise objectives • Align with ERM

Figure 72—Risk Taxonomy

A risk taxonomy provides a clear understanding of terminologies and scales to be used among the stakeholders while discussing and communicating on risk. The risk taxonomy to be used for a risk universe needs to be committed to by all stakeholders.

It also contains methods and procedures applied in the enterprise on risk management (identification, assessment, control, reporting, monitoring, etc.).

	Life Cycle Stage	Internal Stakeholder	External Stakeholder	Description/Stake
Life Cycle and Stakeholders	**Information planning**	ERM committee, risk function		Initiates and drives the implementation and planning.
	Information design	Risk function, all other stakeholders can be potential contributors to the design		• The risk function is to obtain source information from external sources and the needs of the relevant committees/functions. • Based on the input, the risk function designs the most appropriate taxonomy.
	Information build/acquire	Risk function		• Develops a taxonomy tailored to the enterprise needs that is understandable and usable by the stakeholders. • The risk function thus builds the taxonomy based on his/her design and agrees on it with the relevant stakeholders.
	Information use/operate: store, share, use	All stakeholders, in particular: board, ERM committee, risk function, business executive, CIO, CISO	External audit, regulator	To be used by all stakeholders to enable a common view and terminology on risk management applied throughout the enterprise.
	Information monitor	Risk function, internal audit	External audit	• Risk function: Ongoing monitoring on adequacy, completeness and accuracy of information; semi-annual assessment of performance (MEA01) and controls (MEA02) to maintain the information • Internal audit: Annual validation of format and level of contents
	Information dispose	Risk function		To have a single taxonomy; dispose of old overlapping words in case a new one is created.

		Quality Subdimension and Goals	Description—The extent to which information is…	Relevance	Goal—The risk taxonomy…
Goals	**Intrinsic**	**Accuracy**	correct and reliable	Medium	is accurate but not academic
		Objectivity	unbiased, unprejudiced and impartial	High	is precise and clear to all stakeholders
		Believability	regarded as true and credible	High	is understood as single point of truth
		Reputation	regarded as coming from a true and credible source	Low	is tailored to the enterprise
	Contextual and Representational	**Relevancy**	applicable and helpful for the task at hand	Medium	has precise terminology and the interdependencies are understood
		Completeness	not missing and is of sufficient depth and breadth for the task at hand	High	clarifies key areas and relationships as well as procedures to be applied
		Currency	sufficiently up to date for the task at hand	Low	meets the stakeholder's need for information and guidance and frequent updates are unlikely
		Amount of information	appropriate for the task at hand	Low	level of detail (defined during the design) is appropriate to the user of the taxonomy
		Concise representation	compactly represented	Medium	is short and clear
		Consistent representation	presented in the same format	High	agreed-on big-picture is available
		Interpretability	in appropriate languages, symbols, and units, and the definitions are clear	High	is clear to all stakeholders
		Understandability	easily comprehended	High	is applied consistently by stakeholders, showing that the information is easily comprehended
		Manipulation	easy to manipulate and apply to different tasks	Medium	reuses definitions, symbols, etc., frequently throughout the enterprise, which are signs that the information can be applied to different tasks

Figure 72—Risk Taxonomy (cont.)					
	Quality Subdimension and Goals		**Description—The extent to which information is…**	**Relevance**	**Goal—The risk taxonomy…**
Goals	**Security**	**Availability**	available when required, or easily and quickly retrievable	Low	is at all times available to its stakeholders; an unavailability of 24 hours is accepted in case of incidents
		Restricted access	restricted appropriately to authorised parties	Low	is available to all stakeholders

	Attribute	**Description**	**Value**
Good Practice	**Physical**	Information carrier/media	The information carrier for the risk taxonomy needs to be electronic or an information system (e.g., dashboard) and can be a printed document.
	Empiric	Information access channel	The risk taxonomy is accessible through the ERM portal or printed.
	Syntactic	Code/language	The risk taxonomy contains the following subparts:
		Information type	The risk taxonomy contains an understandable list of terminology for the entire enterprise. It also contains a structured overview of risk scales and methods and procedures to be used.
		Information currency	Information should be current for organisational use.
		Information level	The risk taxonomy covers all the terminologies, scales and methods, and procedures that are applicable throughout the entire enterprise.
	Pragmatic	Retention period	The information items are to be retained for as long as the described terminology/scales/methods/relationships are valid for the enterprise (and taking into account local legal retention requirements).
		Information status	The information produced would be current.
		Novelty	The risk taxonomy on definition is novel; however, not a lot of changes are expected during the life cycle or the enterprise.
		Contingency	The risk taxonomy relies on the following information being understood by the user: • Terminology explained • Scales used • Methods and processes explained If guidance is needed, the enterprise risk department should be able to provide extra guidance to the users.
	Social	Context	The risk taxonomy is only meaningful and to be used in a context of ERM.

Link to Other Enablers	
Processes	The risk taxonomy is an <u>output</u> from the management practice: • EDM03.01 Evaluate risk management. • EDM03.02 Direct risk management. The risk taxonomy is an <u>input</u> for the management practice • APO12.01 Collect data. • APO12.02 Analyse risk. • APO12.03 Maintain a risk profile. • APO12.04 Articulate risk. • APO12.05 Define a risk management action portfolio. • APO12.06 Respond to risk.
Organisational Structures	Under the accountability of the risk function, the enterprise risk group is responsible for the setup of the risk taxonomy. All the other stakeholders of the enterprise need to understand and use the taxonomy.
Infrastructure, Applications and Services	The risk taxonomy is manually maintained by the risk function and the enterprise risk group via the risk management portal.
People, Skills and Competencies	The enterprise risk group needs to have a technical understanding, risk expertise and organisational and business awareness to put together the risk taxonomy. The risk taxonomy should be explained in such a way that the user only needs some basic risk management understanding to comprehend the terminology, scales and methods and procedures.
Culture, Ethics and Behaviour	The availability of the risk taxonomy supports the transparency of trends in risk as well as a risk-aware culture.
Principles, Policies and Frameworks	Related principle: • Consistent approach

Figure 73—Business Impact Analysis			
A business impact analysis (BIA) is an activity to develop a common understanding of the business processes that are specific to each business unit and critical to the survival of an enterprise. The result of this activity is a written BIA document.			

Life Cycle and Stakeholders

Life Cycle Stage	Internal Stakeholder	External Stakeholder	Description/Stake
Information planning	Board, COO	External audit, regulator	Determines the business processes in scope and expected analysis.
Information design	Business process owners, business continuity manager, head IT operations, service manager		Prepares the analysis, by identifying the information required to make the analysis.
Information build/acquire	COO, business process owners, business continuity manager, head IT operations, service manager, risk function		Performs the analysis by gathering the identified information and assessing the information to determine the impact on the business processes.
Information use/operate: store, share, use	COO, business process owners, CIO, risk function, head architect, head IT operations, business continuity manager	External audit, regulator	Makes the BIA document available to all the relevant stakeholders.
Information monitor	Business continuity manager, risk function, audit	External audit	Monitors the evolution of the parameters of the analysis and updates the analysis if required.
Information dispose	COO, business process owners, business continuity manager, risk function		Disposes of the BIA document if the information is not relevant anymore to the enterprise.

Goals

Quality Subdimension and Goals		Description—The extent to which information is…	Relevance	Goal—The BIA…
Intrinsic	Accuracy	correct and reliable	High	source information needs to be accurate (confirmed through business executives and needs to be processed, analysed and presented in the risk management and/or business continuity management and/or GRC application, according to fixed rules)
	Objectivity	unbiased, unprejudiced and impartial	High	information is based on verifiable facts and substantiations, using the common business view established throughout the enterprise
	Believability	regarded as true and credible	Medium	document should contain no contradictory information
	Reputation	regarded as coming from a true and credible source	High	source information is collected from business process owners, business executives and other competent/recognised sources

	Figure 73—Business Impact Analysis *(cont.)*			
	Quality Subdimension and Goals	**Description—The extent to which information is…**	**Relevance**	**Goal—The BIA…**
Goals / Contextual and Representational	**Relevancy**	applicable and helpful for the task at hand	High	shall be structured according to the 'syntax' attribute, and it shall be confirmed by the COO and CRO (BAI04) that the information is relevant
	Completeness	not missing and is of sufficient depth and breadth for the task at hand	High	shall cover the entire enterprise, full core business processes/products and the full risk register
	Currency	sufficiently up to date for the task at hand	High	shall be no older than six months
	Amount of information	appropriate for the task at hand	Medium	information relevance is more important than the amount, but without sufficient details the BIA will be hard to follow by stakeholders not involved in the build-up of the analysis
	Concise representation	compactly represented	High	information shall be concisely represented; this is obtained by collecting, processing, analysing, evaluating and assessing the BIA-related data for the entire enterprise
	Consistent representation	presented in the same format	High	shall always be presented according to the current template
	Interpretability	in appropriate languages, symbols, and units, and the definitions are clear	High	will be spread throughout the enterprise to relevant stakeholders, who are not all familiar with the aspects described
	Understandability	easily comprehended	Medium	will be spread throughout the enterprise to relevant stakeholders, who are not all familiar with the aspects described
	Manipulation	easy to manipulate and apply to different tasks	Low	is focused on specific areas within the enterprise and thus will most likely affect only a selected number of tasks
Security	**Availability**	available when required, or easily and quickly retrievable	High	shall at all times be available to its stakeholders; an unavailability is acceptable in case of incidents
	Restricted access	restricted appropriately to authorised parties	High	access is determined by the COO, risk function and business continuity manager, and is restricted as follows: • Write access: COO, business process owners, risk function and business continuity manager • Read access: Entire enterprise

Attribute	**Description**	**Value**
Physical	Information carrier/media	The information carrier for the BIA can be an electronic or printed document.
Empiric	Information access channel	The BIA is accessible through the enterprise management portal or printed at a defined location.
Syntactic	Code/language	The BIA contains the following subparts: • Business continuity-related source data • Risk scenarios • Critical business functions and processes • Vulnerability analysis • Potential disruption threats
Semantic	Information type	Risk report
	Information currency	The BIA handles the most recently available information to assess the impact.
	Information level	The BIA collects data across the entire enterprise and analyses it.

Figure 73—Business Impact Analysis *(cont.)*			
	Attribute	**Description**	**Value**
Good Practice	**Pragmatic**	Retention period	The BIA is to be retained for as long as the data/information over which it reports risk needs to be retained (and taking into account local legal retention requirements).
		Information status	All instances produced over the last six months are operational; older ones are historical data.
		Novelty	The BIA combines several other sources of data that make up a new instance; hence, it is novel data. It is updated on a quarterly basis.
		Contingency	The BIA relies on the following information being available and understood by the user: • Business continuity • Business impact • Risk scenarios • Critical business functions and processes • Vulnerability analysis • Potential disruption threats • Business continuity-related risk events over the entire enterprise • Risk appetite of the enterprise • Risk profile
	Social	Context	The BIA is only meaningful and to be used in a context of enterprise business continuity and risk management.
Link to Other Enablers			
Processes		The BIA is an <u>output</u> from the process activities: • BAI04.02 Assess business impact. • DSS04.02 Maintain a continuity strategy. The BIA is an <u>input</u> for the process activity: • APO12.02 Analyse risk.	
Organisational Structures		The following roles are accountable and responsible for (partially) producing the information: • Accountable: COO • Responsible: Business process owners, business continuity manager, CIO, risk function, head architect, head IT operations	
Infrastructure, Applications and Services		The BIA can be produced by business continuity application or risk management application or GRC application and can be used by service management, monitoring, compliance and audit applications.	
People, Skills and Competencies		The generation and use of a BIA requires basic understanding of business continuity and risk management principles and skills.	
Culture, Ethics and Behaviour		Stakeholders must be educated towards understanding and taking appropriate action on BIAs.	
Principles, Policies and Frameworks		Related principles: • Align with ERM • Balance cost/benefit of IT risk	

177

		Figure 74—Risk Event		
A risk event is something that might happen with an unknown frequency at an unknown future place and/or time, and that will have an impact on the achievement of enterprise objectives.				

	Life Cycle Stage	**Internal Stakeholder**	**External Stakeholder**	**Description/Stake**
Life Cycle and Stakeholders	Information planning	ERM committee, CIO, risk function	External audit, regulator	Determine the internal (For which business process will risk events be monitored?) and external (Which external categories of factors are to be considered?) scope of the risk events.
	Information design	CIO, risk function		Design and detail what information is required when identifying a risk event.
	Information build/acquire	CIO, risk function, business process owners, project management office, CISO, head architect, head development, head IT operations, head IT administration, service manager, information security manager, business continuity manager, privacy officer		Identify in-scope risk events and gather the required information on the risk events.
	Information use/operate: store, Share, Use	Entire enterprise	External audit, regulator	Make use of the identified risk event and take actions accordingly.
	Information monitor	CIO, audit	External audit	Monitor the evolution of the risk event and verify accuracy of information on a regular basis.
	Information dispose	CIO, risk function		Delete the risk event if no longer relevant.

		Quality Subdimension and Goals	**Description—The extent to which information is…**	**Relevance**	**Goal—The risk event…**
Goals	**Intrinsic**	Accuracy	correct and reliable	High	source information needs to be accurate (confirm through audit) and needs to be processed, analysed and presented in the risk management application according to fixed rules
		Objectivity	unbiased, unprejudiced and impartial	High	information is based on verifiable facts and substantiations, using the common risk view established throughout the enterprise
		Believability	regarded as true and credible	Medium	may not be regarded as credible, when dealing with many 'unknowns'. This can be avoided by having many (trustworthy) sources supporting the assumptions. Ranges and variables used for showing the magnitude of issues need to be in a form to which the audience can relate.

Figure 74—Risk Event *(cont.)*				
Quality Subdimension and Goals		**Description—The extent to which information is…**	**Relevance**	**Goal—The risk event…**
Contextual and Representational	**Reputation Relevancy**	is regarded as coming from a true and credible source	High	source information is collected from competent and recognised sources
		applicable and helpful for the task at hand	Medium	information shall be structured according to the 'syntax' attribute and it shall be confirmed by the CIO and risk function (APO12) that the information is relevant
	Completeness	not missing and is of sufficient depth and breadth for the task at hand	High	necessary information will be determined on design, and needs to be aligned with the information needs of the users
	Currency	sufficiently up to date for the task at hand	High	shall be no older than three months
	Amount of information	appropriate for the task at hand	Low	has many 'unknowns' that may make it impossible to collect a lot of information on a risk event
	Concise representation	compactly represented	High	shall be concisely represented; this is obtained by collecting, processing and analysing the risk-related data for the entire enterprise and by detecting and presenting only relevant risk events
	Consistent representation	presented in the same format	Medium	Shall always be presented according to the current template
	Interpretability	in appropriate languages, symbols, and units, and the definitions are clear	Medium	and related information need to be expressed in terms that are widely understood, because the risk events need to be used by a wide range of employees
	Understandability	easily comprehended	Medium	and related information need to be expressed in terms that are widely understood, because the risk events need to be used by a wide range of employees
	Manipulation	easy to manipulate and apply to different tasks	Medium	can have a large impact on the enterprise, and thus needs to be described in such a way that is applicable to the different tasks that might be impacted by the risk event
Security	**Availability**	available when required, or easily and quickly retrievable	Medium	shall at all times be available to its stakeholders; an unavailability of 24 hours is acceptable in case of incidents
	Restricted access	restricted appropriately to authorised parties	High	The access to the risk event is determined by the risk function and CIO, and is restricted as follows: • Write access: Risk function and CIO • Read access: All other stakeholders

*(Row group label at far left: **Goals**)*

	Attribute	Description	Value
Good Practice	**Physical**	Information carrier/media	The information carrier for the risk event can be an electronic or printed document.
	Empiric	Information access channel	The risk event is accessible through the ERM portal or printed at a defined location.
	Syntactic	Code/language	The risk event contains the following subparts: • Risk-related source data • Risk profile relations
	Semantic	Information type	Predefined format in a risk report style summary.
		Information currency	The risk event analyses historical and current data and makes assumptions on future events and states.
		Information level	Data is collected over the entire enterprise, and subsequently processed and analysed to identify risk events.
	Pragmatic	Retention period	The risk event is to be retained for as long as the event over which it reports risk needs to be retained (and taking into account local legal retention requirements).
		Information status	All instances produced over the last six months are operational; older ones are historical data, unless applicability is re-confirmed.
		Novelty	The risk event combines several other sources of data that make up a new instance; hence, it is novel data. It is updated on a weekly basis.
		Contingency	The risk event relies on the following information being available and understood by the user: • Risk-related events over the entire enterprise • Risk appetite of the enterprise • Risk profile • Risk taxonomy in use in the enterprise
	Social	Context	The risk event is only meaningful and to be used in a context of ERM of risk.

Figure 74—Risk Event (cont.)

Link to Other Enablers

Processes	The risk event is an <u>output</u> from the process activity: • APO12.01 Collect data. The risk event is an <u>input</u> for the process activities: • Internal
Organisational Structures	The following roles are accountable and responsible for (partially) producing the information: • Accountable: CIO • Responsible: Risk function, business process owners, project management office, CISO, head architect, head development, head IT operations, head IT administration, service manager, information security manager, business continuity manager, privacy officer
Infrastructure, Applications and Services	This information item can be produced by the risk management application and can be used by service management, monitoring, compliance and audit applications.
People, Skills and Competencies	The generation and use of a risk event requires basic understanding of risk management principles and skills.
Culture, Ethics and Behaviour	Stakeholders must be educated towards determination of risk events.
Principles, Policies and Frameworks	Related principles: • Connect to enterprise objectives • Align with ERM

Figure 75—Risk and Control Activity Matrix (RCAM)			
The risk and control activity matrix (RCAM) is a document that contains identified risk and its ranking, and control activities to respond to the risk, description of their design and assessments of their operating effectiveness.			

Life Cycle and Stakeholders

Life Cycle Stage	Internal Stakeholder	External Stakeholder	Description/Stake
Information planning	ERM committee		Validates and enforces the applicability of the RCAM as a risk management tool.
Information design	Risk function		Identifies the areas/process where an RCAM will be put in place.
Information build/acquire	Risk manager, risk analysts, technical experts, business process owners, compliance		Details the risk and design of the control activities for each process where an RCAM has been set up.
Information use/operate: store, share, use	Business process owners, compliance, internal audit	External audit, regulator	The business process owners will be responsible for making sure the control activities are performed as designed and will resolve any operating effectiveness issues.
Information monitor	Compliance, internal audit, enterprise risk group, ERM committee	External audit, regulator	• Compliance, internal audit and external audit will all monitor the RCAM in the sense that they will verify if the identified risk are indeed relevant, if the design of the control activities is reflecting reality and if they operate as designed. • The ERM committee and regulators will act on any findings from the above parties.
Information dispose	Enterprise risk group, business process owners		Enterprise risk group will ensure that information in the RCAM is still up to date, together with the business process owners, and adapt risk and controls accordingly.

Goals

Quality Subdimension and Goals		Description—The extent to which information is…	Relevance	Goal
Intrinsic	Accuracy	correct and reliable	High	The risk needs to be accurately described to assess if the control objectives are appropriately designed. The control objectives need to be accurately described so there is no room for interpretation in the control operations.
	Objectivity	unbiased, unprejudiced and impartial	High	The risk and control objectives need to be described in an unbiased way, free from the risk appetite of the enterprise.
	Believability	regarded as true and credible	High	The risk identified needs to be regarded as credible related to the process. Otherwise the control objectives covering the risk will not be regarded as believable to mitigate the risk of a process.
	Reputation	regarded as coming from a true and credible source	Low	RCAM source information is collected from workshops and entered into the enterprise risk portal.

	Figure 75—Risk and Control Activity Matrix (RCAM) *(cont.)*			
	Quality Subdimension and Goals	**Description—The extent to which information is…**	**Relevance**	**Goal**
Goals / Contextual and Representational	**Relevancy**	applicable and helpful for the task at hand	Medium	The information required will be detailed in a preapproved template and should be applicable to the specific tasks in scope of the RCAM, but not necessarily unique.
	Completeness	not missing and is of sufficient depth and breadth for the task at hand	High	The controls need to be complete in order to cover all the identified risk in a process.
	Currency	sufficiently up to date for the task at hand	High	The risk and subsequent control objectives and activities need to be current and reflect at all times the current reality.
	Amount of information	appropriate for the task at hand	Medium	The information in the RCAM should provide relevant information to execute the controls and allow a review of the design effectiveness of the RCAM.
	Concise representation	compactly represented	Medium	The information in the RCAM shall always be presented according to the predefined formats/templates, linking a risk to a control with the objective to cover that risk.
	Consistent representation	presented in the same format	High	The information in the RCAM shall always be presented according to the predefined formats/templates, linking a risk to a control with the objective to cover that risk.
	Interpretability	in appropriate languages, symbols, and units, and the definitions are clear	High	The information in the RCAM should leave no room for interpretability. The design of the controls should clearly show that they address the identified risk.
	Understandability	easily comprehended	Medium	The RCAM should be comprehensible for target audience, and could require some risk management knowledge to understand it fully.
	Manipulation	easy to manipulate and apply to different tasks	Low	Controls are specifically designed for a process to mitigate the identified risk of that process.
Security	**Availability**	available when required, or easily and quickly retrievable	Medium	The RCAM shall have medium availability; an unavailability of 24 hours is acceptable.
	Restricted access	restricted appropriately to authorised parties	High	The access to the RCAM is restricted as follows: • Write access: Enterprise risk group • Read access: All other stakeholders
	Attribute	**Description**	**Value**	
Good Practice	**Physical**	Information carrier/media	The information carrier for the RCAM can be an electronic and/or printed document.	
	Empiric	Information access channel	The RCAM is accessible through the ERM portal and is printed out at the RCAM owners.	
	Syntactic	Code/language	The RCAM contains risk, its rankings, risk descriptions, control objectives and control designs with related tolerance levels for operating effectiveness.	
	Semantic	Information type	Standard tabular template	
		Information currency	Information should be current for organisational use.	
		Information level	Detailed to avoid misinterpretation.	
	Pragmatic	Retention period	The RCAMs are retained for maximum one year; then they are reviewed and updated.	
		Information status	The RCAM information is always current or should be updated immediately if not current anymore.	
		Novelty	It is current as the risk, and related controls, are updated in time.	
		Contingency	Information should be clear to the stakeholders and would only require a basic understanding of risk management practices.	
	Social	Context	This RCAM should be used in daily operations to guide the controls that need to be executed in a process to prohibit identified risk from occurring.	

Figure 75—Risk and Control Activity Matrix (RCAM) *(cont.)*	
Link to Other Enablers	
Processes	The RCAM is an <u>output</u> from the process activity: • APO12.05 Define a risk management action portfolio. The RCAM is an <u>input</u> for the process activities: • APO02.02 Assess the current environment, capabilities and performance. • DSS01.01 Perform operational procedures.
Organisational Structures	The following roles are accountable and responsible for (partially) producing the information: • Accountable: CRO, ERM Committee • Responsible: Enterprise risk group, compliance, business process owners • Consulted/Informed: Business internal audit, external audit, regulator (see stakeholders)
Infrastructure, Applications and Services	The RCAM is produced by risk management portal (GRC Tool).
People, Skills and Competencies	The RCAM requires following competencies: • Analytical capability • Lateral thinking (thinking outside of the box) • Risk expertise
Culture, Ethics and Behaviour	The RCAM requires following behaviours: • Recognition of exposure • Ownership of risk is accepted by business • Genuine commitment is obtained and resources assigned for execution of actions • Management proactively monitors risk and action plan progress.
Principles, Policies and Frameworks	Related principles: • Align with ERM • Function as part of daily activities • Consistent approach

Figure 76—Risk Assessment				
Risk assessment is the determination of a quantitative or qualitative exposure linked to a risk scenario. The risk assessment is the result of this effort.				
	Life Cycle Stage	**Internal Stakeholder**	**External Stakeholder**	**Description/Stake**
Life Cycle and Stakeholders	**Information planning**	ERM committee, risk function	External audit, regulator	Identify the relevant risk scenario categories and risk scenarios for the enterprise.
	Information design	Risk function, CIO		Determine what information is required for the risk assessment.
	Information build/acquire	Risk function, CIO, business process owners, project management office, CISO, head architect, head development, head IT operations, head IT administration, service manager, information security manager, business continuity manager, privacy officer		Gather the required information for completing the risk assessment.
	Information use/operate: store, share, use	Entire enterprise	External audit, regulator	Make use of the risk assessment to take actions accordingly.
	Information monitor	CIO, audit		Evaluate at a regular basis the relevance of the information in the risk assessment and the relevance of the risk scenario in general.
	Information dispose	CIO, risk function		Delete the risk assessment if no longer relevant.

	Quality Subdimension and Goals		Description—The extent to which information is…	Relevance	Goal—The risk assessment…
Goals	**Intrinsic**	**Accuracy**	correct and reliable	High	source information needs to be accurate (confirm through audit) and needs to be processed, analysed and presented in the risk management application according to fixed rules
		Objectivity	unbiased, unprejudiced and impartial	High	information is based on verifiable facts and substantiations, using the common risk view established throughout the enterprise
		Believability	regarded as true and credible	Medium	should contain no contradictory information
		Reputation	regarded as coming from a true and credible source	High	source information is collected from competent and recognised sources
	Contextual and Representational	**Relevancy**	applicable and helpful for the task at hand	High	information shall be structured according to the 'syntax' attribute, and it shall be confirmed by the CRO (APO12) that the information is relevant
		Completeness	not missing and is of sufficient depth and breadth for the task at hand	High	shall cover the entire enterprise and the full risk register
		Currency	sufficiently up to date for the task at hand	High	shall be no older than six months
		Amount of information	appropriate for the task at hand	Medium	more relevant information is gathered, the better it allows assessing a risk scenario
		Concise representation	compactly represented	High	information shall be concisely represented; this is obtained by collecting, processing and analysing the risk-related data for the entire enterprise and by detecting and presenting only relevant risk
		Consistent representation	presented in the same format	High	shall always be presented according to the current template
		Interpretability	in appropriate languages, symbols, and units, and the definitions are clear	High	needs to be expressed in terms that are widely understood, as it will be used by a wide range of employees
		Understandability	easily comprehended	High	needs to be expressed in terms that are widely understood, as it will be used by a wide range of employees
		Manipulation	easy to manipulate and apply to different tasks	Medium	should cover a wide range of possible consequences, and could thus apply to/ impact different tasks, but most of the time it remains delineated to identified tasks/activities
	Security	**Availability**	available when required, or easily and quickly retrievable	High	shall at all times be available to its stakeholders; an unavailability of 24 hours is acceptable in case of incidents
		Restricted access	restricted appropriately to authorised parties	High	access to the risk assessment is determined by the risk function, and is restricted as follows: • Write access: Risk function • Read access: All other stakeholders

Figure 76—Risk Assessment (cont.)

	Attribute	Description	Value
Figure 76—Risk Assessment (cont.)			
Good Practice	**Physical**	Information carrier/media	The information carrier for the risk assessment can be an electronic or printed document.
	Empiric	Information access channel	The risk assessment is accessible through the ERM portal.
	Syntactic	Code/language	The risk assessment contains the following subparts: • Risk-related source data • Risk profile relations • Risk analysis • Third-party risk analysis
	Semantic	Information type	Risk assessment and scenario template style
		Information currency	The risk assessment contains historical and current data.
		Information level	The risk assessment collects data over the entire enterprise, processes and analyses the data and represents relevant data fully in the risk assessment.
	Pragmatic	Retention period	The risk assessment is to be retained for as long as the data/information over which it reports risk needs to be retained (and taking into account local legal retention requirements).
		Information status	All instances produced over the last 12 months are operational, older ones are historical data.
		Novelty	The risk assessment combines several other sources of data that make up a new instance; hence, it is novel data.
		Contingency	The risk assessment relies on the following information being available and understood by the user: • Risk-related events over the entire enterprise • Risk appetite of the enterprise • Risk analysis • Evaluation of risk management activities • Evaluations of potential threats • Risk profile • Risk taxonomy in use in the enterprise
	Social	Context	The risk assessment is only meaningful and to be used in a context of ERM.

Link to Other Enablers

Processes	The risk assessment is an <u>output</u> from the process activities: • APO02.05 Define the strategic plan and road map. • APO12.02 Analyse risk. • APO12.04 Articulate risk. • BAI01.10 Manage programme and project risk. The risk assessment is an <u>input</u> for the process activities: • EDM03.03 Monitor risk management • APO01.03 Maintain the enablers of the management system. • APO02.02 Assess the current environment, capabilities and performance. • APO05.01 Establish the target investment mix. • APO10.04 Manage supplier risk. • APO12.01 Collect data. • BAI01.10 Manage programme and project risk. • MEA02.01 Monitor internal controls.
Organisational Structures	The following roles are accountable and responsible for partially producing the information: • Accountable: Risk function • Responsible: CIO, business process owners, project management office, CISO, head architect, head development, head IT operations, head IT administration, service manager, information security manager, business continuity manager, privacy officer
Infrastructure, Applications and Services	The risk assessment can be produced by risk management application and can be used by project and portfolio management, service management, monitoring, compliance and audit applications.
People, Skills and Competencies	The generation and use of a risk assessment requires basic understanding of risk management principles and skills.
Culture, Ethics and Behaviour	Stakeholders must be educated about the meaning of and how to use a risk assessment.
Principles, Policies and Frameworks	Related principles: • Connect to enterprise objectives • Promote fair and open communication • Establish tone at the top and accountability

Page intentionally left blank

B.6. Enabler: Services, Infrastructure and Applications

For each of the following services, this section details the description, good practices (architecture principles and viewpoints and service level considerations) and the related stakeholders:
• **Figure 77—Programme/Project Risk Advisory Services**
• **Figure 78—Incident Management Services**
• **Figure 79—Architecture Advisory Services**
• **Figure 80—Risk Intelligence Services**
• **Figure 81—Risk Management Services**
• **Figure 82—Crisis Management Services**

Figure 77—Programme/Project Risk Advisory Services		
Description	**Goal/Purpose**	**Benefit**
The set of people, processes, and technology that helps to ensure new or changing business strategies, products, processes or technologies do not introduce unacceptable levels of risk to the enterprise.	Help the enterprise achieve/maintain an optimised level of risk.	By providing this service before changes are implemented, the enterprise will be proactive in managing risk and can avoid costly re-engineering to mitigate avoidable loss exposure.
Good Practices		
Description	**Good Practices**	
Buy	Enterprises can choose to outsource this service or build in-house capabilities as appropriate for their culture and resource.	
Use	• Establish a triage process to help ensure that the service is applied to programmes/projects with the highest potential for introducing new or increased risk to the enterprise. • Establish a governance process to ensure that programmes/projects do not inappropriately circumvent the service. For example, establishing a process that only releases funds to programmes/projects after risk management milestones are achieved (e.g., earned value management).	
Stakeholders		
Stakeholders	**Description**	
Users	Business executives, programme management office, CIO, risk function, CTO, CISO	
Sponsors	CIO, business strategy office, audit, compliance	

Figure 78—Incident Management Services		
Description	**Goal/Purpose**	**Benefit**
The set of people, processes, and technology that help to minimise losses that materialise from incidents.	Minimise losses that materialise from incidents.	Lower loss magnitudes
Good Practices		
Description	**Good Practices**	
Buy	Enterprises can choose to outsource this service or build in-house capabilities as appropriate for their culture and resources. • Enterprises should ensure that the appropriate stakeholders are engaged in the incident management team (e.g., legal, privacy, HR). • The service should include processes that accurately document steps taken to resolve incidents as they take place, for later reference (e.g., in legal defence proceedings). • The service should leverage competent forensic resources as appropriate.	
Use	The service should document losses and expenses that are incurred to provide empirical evidence for later risk analyses.	
Stakeholders		
Stakeholders	**Description**	
Users	Business executives, CIO, risk function, CTO, CISO, physical security management	
Sponsors	CIO, audit, privacy, legal, compliance	

Figure 79—Architecture Advisory Services

Description	Goal/Purpose	Benefit
The set of people, processes, and technology that help to ensure new or changing business or technology architecture do not introduce unacceptable levels of risk to the enterprise.	Help the enterprise architecture to support an optimised level of risk.	Reduces the potential for changes in architecture to introduce unintended and/or unacceptable levels of risk.

Good Practices	
Description	**Good Practices**
Buy	Enterprises can choose to outsource this service or build in-house capabilities as appropriate for their culture and resources.
Use	Enterprises should ensure that they have effective sources of intelligence regarding tactical and strategic (emerging) changes in the threat landscape and technologies.

Stakeholders	
Stakeholders	**Description**
Users	CTO, CIO, business executives, programme management office, risk function, CISO
Sponsors	CIO, CTO, CISO, business strategy office, audit, compliance

Figure 80—Risk Intelligence Services

Description	Goal/Purpose	Benefit
The set of people, processes, and technology that help to ensure accurate and timely threat, vulnerability and asset intelligence is available. This intelligence needs to include both tactical and strategic (longer time horizon) information.	Help analysts and decision makers make better-informed decisions.	• The enterprise is less likely to overlook, or undervalue key components of the risk environment. • The enterprise is able to be more proactive in its decision making.

Good Practices	
Description	**Good Practices**
Buy	Enterprises can choose to outsource this service or build in-house capabilities as appropriate for their culture and resources.
Use	The enterprise needs to identify reliable sources of intelligence and integrate these sources into the analysis and decision-making process.

Stakeholders	
Stakeholders	**Description**
Users	Business executives, programme management office, CIO, risk function, CTO, CISO
Sponsors	CIO, business strategy office, audit, compliance

Figure 81—Risk Management Services

Description	Goal/Purpose	Benefit
The set of people, processes, and technology that help to ensure the enterprise build/ maintain a cost-effective programme for managing risk over time.	To provide subject matter expertise to key risk management stakeholders.	• Helps to ensure the ERM programme is comprehensive, consistent, and effectively integrated into business lines and their processes. • Provides a holistic view on risk throughout the enterprise.

Good Practices	
Description	**Good Practices**
Buy	Enterprises can choose to outsource this service or build in-house capabilities as appropriate for their culture and resources.
Use	Adopt or develop a standard framework for risk management that will be used throughout the enterprise.

Stakeholders	
Stakeholders	**Description**
Users	All business line executives, programme management office, CIO, risk function CTO, CISO
Sponsors	Board, CEO, CRO, audit

Figure 82—Crisis Management Services		
Description	**Goal/Purpose**	**Benefit**
The set of people, organisations, processes and technology that help to respond to any type of crisis including those requiring the activation of the BCP.	• Define the general crisis management principles applicable: organisation of arrangements, roles of the actors involved. • Specify the procedures and processes to be implemented, the logistic resources to be used and the principles of maintenance in operational status for the services.	Reduce the impact of major adverse events and avoid issues such as: • Information redundancy, retention and relevancy • Inefficiency of the enterprise due to inappropriate logistics • Multiple decision centres and no respect of position and responsibility • Poor decisions due to a lack of communication channels towards management
Good Practices		
Description	**Good Practices**	
Buy	Enterprises can choose to outsource this service partially or build in-house capabilities as appropriate for their culture and resources.	
Use	The following items should be foreseen by the crisis management services: • Physical resources: assets, fixtures and fittings for crisis management rooms • Information resources: memos, directories, technical and personal data • Technical resources: telecommunication technology, secured remote platforms • Human resource expertise for crisis management • Crisis management teams and experts composition, function and contact details	
Stakeholders		
Stakeholders	**Description**	
Users	Business executives, CIO, risk function, CTO, CISO, physical security management, business continuity manager	
Sponsors	Business continuity manager, risk function, privacy, legal and compliance	

Page intentionally left blank

B.7. Enabler: People, Skills and Competencies

Key senior management roles are described in detail as part of the Organisational Structures enabler and are, therefore, not elaborated on in this section.
• ERM committee
• Enterprise risk group
• Risk function
• Audit department
• Compliance department

The following section further outlines the skills and competencies of two specific risk roles with risk management responsibilities:
• **Figure 83—Risk Manager**
• **Figure 84—Risk Practitioner**

Figure 83—Risk Manager
Risk managers are responsible for the successful implementation and monitoring of the risk strategy and framework. Risk managers engage with stakeholders to ensure that risk management processes are understood, resourced and implemented, and support business goals. Risk outcomes are reported to the CRO for incorporation into overall risk profiles and risk issues.
The role works with business management to ensure that the overall information technology risk function effectively supports strategic goals. The risk manager collaborates with audit/business segment/corporate risk to address issues with plausible action plans and target dates. This role acts as the central point for receipt and distribution of important risk information for information technology and reciprocates the flow of information back to corporate risk management. The risk manager ensures that information technology adheres to corporate and business unit policies and procedures. The role must be aware of and keep abreast of technology risk associated with the enterprise. The role may or may not have managerial responsibility.
This figure describes the typical experience, education and qualifications for this specific role. These should not be considered strict requirements, but guidance that can be used as input, e.g., when detailing job descriptions.

Experience, Education and Qualifications	
Requirement	**Description**
Experience	• Adequate experience in managing and governing business risk and/or operations • Experience in communication of risk to executive management and/or board
Education	Degree in management information systems with experience in IT, finance, economics, business or engineering
Typical qualifications and certifications	CISA, CISM, CRISC, CISSP, CPA
Knowledge, Technical Skills and Behavioural Skills	
Knowledge	• Have a deep knowledge of the enterprise and the IT systems that support the business functions as well as be aware of the contextual factors that influence them • Solid knowledge of risk methodologies, commonly used risk standards and risk best practices
Technical skills	Have knowledge of the technical side of IT systems supporting the business functions
Behavioural skills	• Leadership • Communication • Influencing • Patience • Escalation

Figure 84—Risk Analyst

The risk analyst is responsible for:
• Executing the overall risk assessment process in the enterprise
• Identifying and analysing the areas of potential risk that are threatening assets and the achievement of the organisational objectives
• Provide specific evaluation of risk scenarios by considering the business and the technical perspective
• Reports their findings to the risk manager or CRO.

This table describes the typical experience, education and qualifications for this specific role. These should not be considered strict requirements, but guidance that can be used as input, e.g., when detailing job descriptions.

Experience, Education and Qualifications	
Requirement	**Description**
Experience	• Adequate relevant experience in business administration or IT • Have a consistent knowledge in systems architecture, infrastructure, security and applications
Education	• Bachelor's degree in financial analysis, IT, engineer, systems analyst • Master's degree in related discipline, e.g., mathematics, statistics
Typical qualifications and certifications	CISM, CRISC, CISSP, FAIR
Knowledge, Technical Skills and Behavioural Skills	
Knowledge	• Knowledge on risk methodologies, commonly used risk standards, risk good practices, and quantitative and qualitative risk analysis • Consistent knowledge on business processes and their relationship to technology • Use of risk assessment tools and techniques
Technical skills	• Profound IT and business functioning understanding and a profound understanding of IT domains, threats, assets • Analytical capability, with desirable knowledge of statistical analysis and probabilities
Behavioural skills	• Communication skills • Presentation skills • Peer reviews • Decision making • Work delegation to technical experts

APPENDIX C
CORE COBIT 5 RISK MANAGEMENT PROCESSES

This appendix contains more detailed guidance on the core risk management processes that were identified in section 2B, chapter 1.

Figure 85—Core Risk Management Processes	
COBIT 5 Process Identification	**Justification**
EDM03 Ensure Risk Optimisation	Covers the articulation and communication of the enterprise's risk appetite and tolerance and ensures identification and management of the risk related to the enterprise value in the use of IT.
APO12 Manage Risk	Covers the identification, assessment and reduction activities of risk.

For each process, the following information is provided:
• Process description and process purpose statement (identical to *COBIT 5 Enabling Processes*)
• Process goals and metrics
• Process practices (identical to *COBIT 5 Enabling Processes*), input and outputs
• Process activities with an additional level of detailed activities included. This additional, risk-specific information is not included in *COBIT 5: Enabling Processes*.

Page intentionally left blank

EDM03 Risk-specific Ensure Risk Optimisation	Area: Governance Domain: Evaluate, Direct and Monitor

COBIT 5 Process Description
Ensure that the enterprise's risk appetite and tolerance are understood, articulated and communicated, and that risk to enterprise value related to the use of IT is identified and managed.

COBIT 5 Process Purpose Statement
Ensure that IT-related enterprise risk does not exceed risk appetite and risk tolerance, the impact of IT risk to enterprise value is identified and managed, and the potential for compliance failures is minimized.

EDM03 Risk-specific Process Goals and Metrics

Risk-specific goals and metrics are not relevant for this practice. The generic COBIT 5 goals and metrics can be used as further guidance.

EDM03 Risk-specific Process Practices, Inputs/Outputs and Activities

Governance Practice	Risk-specific Inputs (in Addition to COBIT 5 Inputs)		Risk-specific Outputs (in Addition to COBIT 5 Outputs)	
	From	Description	Description	To
EDM03.1 Evaluate risk management. Continually examine and make judgement on the effect of risk on the current and future use of IT in the enterprise. Consider whether the enterprise's risk appetite is appropriate and that risk to enterprise value related to the use of IT is identified and managed.	Risk-specific inputs and outputs are not relevant for this practice. The generic COBIT 5 inputs and outputs can be used as further guidance.			

Risk-specific Activities (in Addition to COBIT 5 Activities)

1. Determine the level of IT-related risk the enterprise is willing to take to meet its objectives (risk appetite).
   1.1 Perform enterprise IT risk assessment.
   1.2 Sponsor workshops with business management to discuss the broad amount of risk that the enterprise is willing to accept in pursuit of its objectives (risk appetite).
   1.3 Help business managers understand IT risk in the context of scenarios that affect their business and the objectives that matter most in their daily lives.
   1.4 Take a top-down, end-to-end look at business services and processes and identify the major points of IT support. Identify where value is generated and needs to be protected and sustained.
   1.5 Identify IT-related events and conditions that may jeopardise value, affect enterprise performance and execution of critical business activities within acceptable bounds, or otherwise affect enterprise objectives. Map them to a business-driven hierarchy of risk categories and subcategories (IT risk domains) derived from high-level IT risk scenarios.
   1.6 Break up IT risk by lines of business, product, service and process. Identify potential cascading and coincidental threat types and the probable effect of risk concentration and correlation across silos.
   1.7 Understand how IT capabilities contribute to the enterprise's ability to add value and withstand loss. Compare management's perception of the importance of IT capabilities to their current state.
   1.8 Consider how IT strategies, change initiatives and external requirements may affect the risk profile.
   1.9 Identify risk focus areas, scenarios, dependencies, risk factors and measurements of risk that require management attention and further examination and development.

EDM03 Risk-specific Process Practices, Inputs/Outputs and Activities *(cont.)*
Risk-specific Activities (in Addition to COBIT 5 Activities) *(cont.)*

2. Evaluate and approve proposed IT risk tolerance thresholds against the enterprise's acceptable risk and opportunity levels.
   2.1 Establish the amount of IT-related risk a line of business, product, service, process, etc., is willing to take to meet its objectives (risk appetite).
   2.2 Express limits in measures similar to the underlying business objectives and against acceptable and unacceptable business impacts.
   2.3 Consider any trade-offs that may be required to achieve key objectives in the context of risk-return balance.
   2.4 Propose limits and measures in the context of IT benefit/value enablement, IT programme and project delivery, and IT operations and service delivery, and over multiple time horizons.
   2.5 Evaluate proposed IT risk tolerance thresholds against the enterprise's acceptable risk and opportunity levels, taking into account the results of enterprise IT risk assessment and trade-offs required to achieve key objectives in the context of risk-return balance.
   2.6 Consider the potential effects of IT risk concentration and correlation across lines of business, product, service and process, and determine whether any unit-specific tolerance thresholds should be applied to all business lines.
   2.7 Define the types of events (internal or external) and changes to business environments or technologies that may necessitate a modification to the IT risk tolerance.
   2.8 Approve IT risk tolerance thresholds.

3. Determine the extent of alignment of the IT risk strategy to enterprise risk strategy.
   3.1 Codify IT risk appetite and tolerance into policy at all levels across the enterprise.
   3.2 Recognise that IT risk is inherent to enterprise objectives and document how much IT risk is desired and allowed in pursuit of those objectives.
   3.3 Document risk management principles, risk focus areas and key measurements.
   3.4 Recommend adjustments to the IT risk policy based on changing risk conditions and emerging threats.
   3.5 Align operational policy and standards statements with risk tolerance.
   3.6 Perform periodic or triggered reviews of operational policy and standards against IT risk policy and tolerance. Where there are gaps, set target dates based on acceptable risk exposure time limits and required resources.
   3.7 Propose adjustments to risk tolerance instead of modifying established and effective operational policy and standards.

4. Proactively evaluate IT risk factors in advance of pending strategic enterprise decisions and ensure that risk-aware enterprise decisions are made.
   4.1 Determine the risk and performance levels within the portfolio of IT applications as compared to the value of the business processes they enable, or opportunities to rebalance the enterprise portfolio based on risk, return and anticipated changes to the IT environment.
   4.2 Help business management consider the effect that IT risk and existing risk management capacity (controls, capabilities, resources) will have on business decisions and the effect business decisions may have on existing IT risk exposure and IT risk management capacity going forward.
   4.3 Help business management understand IT risk based on various portfolio views (e.g., business unit, product, process) and weigh the impact that proposed IT investments will have on the overall risk profile of the enterprise (increase or reduce risk).
   4.4 Stress that as a condition of approval of business decisions, cost and opportunity must be weighed against an estimated net change of the IT risk exposure (i.e., impact).

5. Determine that IT use is subject to appropriate risk assessment and evaluation, as described in relevant international and national standards.
   5.1 Provide input for management decision makers on the proposed IT risk analysis approach.
   5.2 Illustrate how risk analysis results can benefit major decisions.
   5.3 Describe what level of quality decision makers should expect, how to interpret risk analysis reports, definitions of key terms (e.g., risk probabilities, degree of error, risk factors), and the limitations of measurements and estimates based on incomplete data.
   5.4 Identify gaps with enterprise risk expectations.
   5.5 Determine whether the risk analysis report provides sufficient information to understand the risk issues and, if needed, to evaluate risk response options. Note its limitations for the decision(s) at hand.

6. Evaluate risk management activities to ensure alignment with the enterprise's capacity for IT-related loss and leadership's tolerance of it.
   6.1 Determine how IT-related risk management is to be defined in the context of protecting and sustaining a given business process or business activity.
   6.2 Adopt and align with the existing enterprise framework for business risk.
   6.3 Integrate any IT specifics into one enterprise approach. Understand the enterprise's risk goals and objectives and the mix of competing business risk issues and resource limitations.
   6.4 Determine how IT risk management is to be approached in the context of the enterprise's risk universe and other enterprise risk types.
   6.5 Define the IT department's role in operational risk management activities based on the extent of the business's dependency on IT and related physical infrastructure in achieving financial, operational and customer-satisfaction objectives.
   6.6 Co-ordinate risk assessment activities and perform integrated reporting.
   6.7 Co-ordinate risk and issue classification; risk rating scales and hierarchies for risk-based policies,

	Risk-specific Inputs (in Addition to COBIT 5 Inputs)		Risk-specific Outputs (in Addition to COBIT 5 Outputs)	
EDM03 Risk-specific Process Practices, Inputs/Outputs, Activities and Detailed Risk Activities *(cont.)*				
Governance Practice	**From**	**Description**	**Description**	**To**
EDM03.2 Direct risk management. Direct the establishment of risk management practices to provide reasonable assurance that IT risk management practices are appropriate to ensure that the actual IT risk does not exceed the board's risk appetite.	Risk-specific inputs and outputs are not relevant for this practice. The generic COBIT 5 inputs and outputs can be used as further guidance.			

Risk-specific Activities (in Addition to COBIT 5 Activities)

1. Promote an IT risk-aware culture and empower the enterprise to proactively identify IT risk, opportunity and potential business impacts.
  1.1 Encourage employees to address IT risk issues before serious escalation is required.
  1.2 Train business and IT staff on threats, impacts and the enterprise's planned responses to specific risk events.
  1.3 Communicate the 'why you should care' message for risk focus areas, and explain how to take risk-aware actions for situations not specified in policies.
  1.4 Walk through scenarios for areas not directly covered by policy, and reinforce expectations for understanding general policy direction and using good judgement.
  1.5 Demonstrate an attitude that encourages discussion and acceptance of the appropriate amount of risk, i.e., be positive about promoting a risk culture appropriate for IT and aligned with enterprise risk-aware culture.

2. Direct the integration of the IT risk strategy and operations with the enterprise strategic risk decisions and operations.
  2.1 Organise existing IT risk management methods required to: 1) understand the business context for IT (e.g., business activity IT dependency analysis, scenario analysis); 2) identify IT risk (e.g., data model, escalation pathways); 3) govern IT risk; and 4) manage IT risk (e.g., select the right KRIs for the right business performance targets and define escalation procedures).
  2.2 Understand ERM expectations, activities and methods that are relevant to IT risk management.
  2.3 Identify gaps and specific IT risk management practices that should be updated or created to meet ERM expectations.
  2.4 Identify enterprise risk activities that should be added or updated to fully consider IT risk.
  2.5 Identity what other functions do, or need to do, in support of the enterprise's objectives and management of IT risk.
  2.6 Prioritise and track efforts to close the gaps between IT risk and ERM, and improve effectiveness and efficiency (e.g., optimise controls, streamline risk assessment, co-ordinate KRIs and escalation triggers, integrate reporting).

3. Direct the development of risk communication plans (covering all levels of the enterprise) as well as risk action plans.
  3.1 Establish and maintain a risk communication plan that covers IT risk policy, responsibilities, accountabilities and the risk landscape. Feature filters in the plan so it is clear, concise, useful and directed to the right audience.
  3.2 Perform frequent and regular communication between IT management and business leadership regarding IT risk issue status, concerns and exposures.
  3.3 Base business and IT management communications on a predefined approach with the following objectives:
    • Align IT risk communication with enterprise risk terms.
    • Consistently prioritise IT risk issues in a manner that aligns with the enterprise definition of business risk.
    • Express IT risk in business strategic and operational terms.
    • Clearly communicate how adverse IT-related events may affect business objectives.
    • Enable senior managers and IT executives to understand the actual amount of IT risk to help steer the right resources to respond to IT risk in line with appetite and tolerance.

4. Direct that the appropriate mechanisms are in place to respond quickly to changing risk and report immediately to appropriate levels of management, supported by agreed principles of escalation (what to report, when, where and how).
  4.1 Specify those accountable and responsible for the management of IT risk across the enterprise. For the top-level executive with overall accountability for IT risk, set a performance expectation to incorporate risk awareness in the culture.
  4.2 Ensure that there are structures in place (e.g., enterprise risk committee, IT risk council, IT risk officer) to involve the business with risk-return-aware decisions and day-to-day operations.
  4.3 Create a distinction amongst the roles of business units (who own and manage the risk on a day-to-day basis), risk control functions (who offer subject matter expert evaluation and advice) and internal audit (who provide independent assurance).
  4.4 Assign roles for managing specific IT risk domains (e.g., system capacity, IT staffing, IT programme selection). Assign each domain a level of criticality based on risk/reward. As required, assign additional risk management responsibilities (e.g., system-specific) at lower levels.
  4.5 Examine the portfolio of risk response activities to identify those with a greater probable impact on overall risk reduction. Quantify the overall expected effect on the probable frequency and magnitude of related risk scenarios through the planned application of controls, capability and resources.
  4.6 Based on dimensions such as current risk level and effectiveness/cost ratio, classify and balance responses (e.g., quick wins, opportunities, deferred efforts) with those that may need a business case.
  4.7 Emphasise specific projects with relatively greater odds to:
    • Reduce risk concentrations (e.g., improvements to architecture, separation of operational units and systems)
    • Implement controls that directly address multiple risk types and are cost-effective
    • Implement controls that improve process effectiveness and prevent excessive risk taking
  4.8 Record the rationale, constraints and how the decision is driving changes to published policy, operational controls, capabilities, resource deployments and communication plans. When applicable, record the rationale for exceeding or falling below risk appetite and tolerance.

EDM03 Risk-specific Process Practices, Inputs/Outputs, Activities and Detailed Risk Activities *(cont.)*

Risk-specific Activities (in Addition to COBIT 5 Activities) *(cont.)*

5. Direct that risk, opportunities, issues and concerns may be identified and reported by anyone at any time. Risk should be managed in accordance with published policies and procedures and escalated to the relevant decision makers.

  5.1 Using the established IT risk tolerance thresholds as a guide, decide whether to accept the remaining risk exposure level.

  5.2 Consider relevant information from risk analysis reports such as loss probabilities and ranges, risk response options, cost/benefit expectations, and the potential effects of risk aggregation. Discuss with impacted business process owners and together examine the risk-return ratios, and determine where to spend the risk budget on 'known' risk to allow acceptance of the unknown risk.

  5.3 Obtain business agreement on risk acceptance or, if no acceptance, the appropriate risk response requirements. Document how risk was considered in the decision and the rationale for any exceptions to risk tolerance (e.g., significant strategic business opportunity).

  5.4 Ensure that risk acceptance decisions and risk response requirements are communicated across organisational lines in accordance with established enterprise risk and corporate governance policies and procedures.

6. Identify key goals and metrics of risk governance and management processes to be monitored, and approve the approaches, methods, techniques and processes for capturing and reporting the measurement information.

  6.1 Establish performance measurements and reporting processes with appropriate levels of recognition, approval, incentives and sanctions.

  6.2 Identify business managers with authority to address IT risk issues across IT benefit/value enablement, IT programme and project delivery, and IT operations and service delivery. Set expectations for these managers to champion policies, standards, controls and compliance monitoring activities (e.g., establishment and monitoring of KRIs).

  6.3 Set and evaluate performance targets based on risk-return-aware decision making (e.g., the ability of managers to integrate and balance performance management with risk management across their scope of authority).

Governance Practice	Risk-specific Inputs (in Addition to COBIT 5 Inputs)		Risk-specific Outputs (in Addition to COBIT 5 Outputs)	
	From	Description	Description	To
EDM03.3 Monitor risk management. Monitor the key goals and metrics of the risk management processes and establish how deviations or problems will be identified, tracked and reported for remediation.	Risk-specific inputs and outputs are not relevant for this practice. The generic COBIT 5 inputs and outputs can be used as further guidance.			

Risk-specific Activities (in Addition to COBIT 5 Activities)

Risk-specific guidance is not relevant for this practice. The generic COBIT 5 activities can be used as further guidance.

APO12 Manage Risk	Area: Management Domain: Align, Plan and Organise

COBIT 5 Process Description
Continually identify, assess and reduce IT-related risk within levels of tolerance set by enterprise executive management.

COBIT 5 Process Purpose Statement
Integrate the management of IT-related enterprise risk with overall ERM, and balance the costs and benefits of managing IT-related enterprise risk.

APO12 Risk-specific Process Goals and Metrics

Risk-specific goals and metrics are not relevant for this practice. The generic COBIT 5 goals and metrics can be used as further guidance.

APO12 Risk-specific Process Practices, Inputs/Outputs, Activities and Detailed Activities

Management Practice	Risk-specific Inputs (in Addition to COBIT 5 Inputs)		Risk-specific Outputs (in Addition to COBIT 5 Outputs)	
	From	Description	Description	To
APO12.01 Collect data. Identify and collect relevant data to enable effective IT-related risk identification, analysis and reporting.	Risk-specific inputs and outputs are not relevant for this practice. The generic COBIT 5 inputs and outputs can be used as further guidance.			

Risk-specific Activities (in Addition to COBIT 5 Activities)

1. Establish and maintain a method for the collection, classification and analysis of IT risk-related data, accommodating multiple types of events, multiple categories of IT risk and multiple risk factors.
   1.1 Establish and maintain a model for the collection, classification and analysis of IT risk data.
   1.2 Accommodate multiple types of events and multiple categories of IT risk.
   1.3 Include filters and views to help determine how specific risk factors may affect risk.
   1.4 The model should support the measurement and assessment of risk attributes across IT risk domains and provide useful data for setting incentives for a risk-aware culture.

2. Record relevant data on the enterprise's internal and external operating environment that could play a significant role in the management of IT risk.
   2.1 Record data on the enterprise's operating environment that could play a significant role in the management of IT risk.
   2.2 Consult sources within the business, legal department, audit, compliance and office of the CIO.
   2.3 Cover major revenue streams, external IT systems, product liability, the regulatory landscape, competition within the industry, IT trends, competitor alignment with key benchmarks, relative maturity in key business and IT capabilities, and geopolitical issues.
   2.4 Survey and organise the historical IT risk data and loss experience of industry peers through industry-based event logs, databases and industry agreements for common event disclosure.

3. Survey and analyse the historical IT risk data and loss experience from externally available data and trends, industry peers through industry-based event logs, databases, and industry agreements for common event disclosure.
   3.1 Per the data collection model, record data on risk events that have caused or may cause impacts to IT benefit/value enablement, IT programme and project delivery, and/or IT operations and service delivery.
   3.2 Capture relevant data from related issues, incidents, problems and investigations.

4. Record data on risk events that have caused or may cause impacts to IT benefit/value enablement, IT programme and project delivery, and/or IT operations and service delivery. Capture relevant data from related issues, incidents, problems and investigations.
   4.1 Organise the collected data and highlight contributing factors.
   4.2 Determine what specific conditions existed or did not exist when risk events were experienced and how the conditions may have affected event frequency and magnitude of loss.
   4.3 Determine common contributing factors across multiple events. Perform periodic event and risk-factor analysis to identify new or emerging risk issues and to gain an understanding of the associated internal and external risk factors.

AP012 Risk-specific Process Practices, Inputs/Outputs, Activities and Detailed Activities *(cont.)*				
	Risk-specific Inputs (in Addition to COBIT 5 Inputs)		**Risk-specific Outputs (in Addition to COBIT 5 Outputs)**	
Management Practice	**From**	**Description**	**Description**	**To**
AP012.02 Analyse risk. Develop useful information to support risk decisions that take into account the business relevance of risk factors.	Risk-specific inputs and outputs are not relevant for this practice. The generic COBIT 5 inputs and outputs can be used as further guidance.			

Risk-specific Activities (in Addition to COBIT 5 Activities)
1. Define the appropriate breadth and depth of risk analysis efforts considering all risk factors and the business criticality of assets. Set the risk analysis scope after performing a cost/benefit analysis. 1.1 Define IT risk analysis scope. Decide on the expected breadth and depth of risk analysis efforts. Consider a wide range of scope options. 1.2 Map in relevant risk factors and the business criticality of in-scope assets/resources and triggers. 1.3 Aim for optimal value from risk analysis efforts by favouring scope based on productive processes and products of the business over internal structures not directly related to business outcomes. 1.4 Set the risk analysis scope after a consideration of business criticality, the cost of measurement vs. the expected value of information and reduction in uncertainty, and any overarching regulatory requirements.
2. Build and regularly update IT risk scenarios, including compound scenarios of cascading and/or coincidental threat types, and develop expectations for specific control activities, capabilities to detect and other response measures. 2.1 Estimate IT risk. Estimate the probable frequency and probable magnitude of loss or gain associated with IT risk scenarios as influenced by applicable risk factors. 2.2 Estimate the maximum amount of damage that could be suffered or opportunity that could be gained. 2.3 Consider compound scenarios of cascading and/or coincidental threat types. 2.4 Based on the most important scenarios, develop expectations for specific controls, capability to detect and other response measures. 2.5 Evaluate known operational controls and their effect on probable frequency, and probable magnitude and applicable risk factors. 2.6 Estimate residual risk exposure levels and compare to acceptable risk tolerance to identify exposures that may require a risk response.
3. Estimate the frequency and magnitude of loss or gain associated with IT risk scenarios. Take into account all applicable risk factors, evaluate known operational controls and estimate residual risk levels. 3.1 Identify risk response options. Examine the range of risk response options, such as avoid, reduce/mitigate, transfer/share, accept and exploit/seize. 3.2 Document the rationale and potential trade-offs across the range. 3.3 Specify high-level requirements for projects or programmes that, based on risk tolerance, will mitigate risk to acceptable levels; identify costs, benefits and responsibility for project execution. 3.4 Develop requirements and expectations for material controls at the most appropriate points, or where they are expected to be rolled up to give meaningful visibility.
4. Compare residual risk to acceptable risk tolerance and identify exposures that may require a risk response. 4.1 Perform a peer review of IT risk analysis. Perform a peer review of the risk analysis results before sending them to management for approval and use in decision making. 4.2 Confirm that the analysis is documented in line with enterprise requirements. 4.3 Review the basis for the estimates of loss/gain probabilities and ranges. 4.4 Verify that any human estimators were properly calibrated beforehand and look for evidence of 'gaming the system', i.e., a conscious or otherwise suspect choice of inputs that may result in a desired or expected outcome. 4.5 Verify that the experience level and credentials of the analyst were appropriate for the scope and complexity of the review. 4.6 Provide an opinion on whether the expected reduction in uncertainty was achieved and whether the value of information gained exceeded the cost of measurement.

APO12 Risk-specific Process Practices, Inputs/Outputs, Activities and Detailed Activities *(cont.)*				
	Risk-specific Inputs (in Addition to COBIT 5 Inputs)		Risk-specific Outputs (in Addition to COBIT 5 Outputs)	
Management Practice	From	Description	Description	To
APO12.03 Maintain a risk profile. Maintain an inventory of known risk and risk attributes (including expected frequency, potential impact and responses) and of related resources, capabilities and current control activities.	Risk-specific inputs and outputs are not relevant for this practice. The generic COBIT 5 inputs and outputs can be used as further guidance.			

Risk-specific Activities (in Addition to COBIT 5 Activities)
1. Inventory business processes, including supporting personnel, applications, infrastructure, facilities, critical manual records, vendors, suppliers and outsourcers, and document the dependency on IT service management processes and IT infrastructure resources.   1.1 Understand the dependency of key business activities on IT service management processes and IT infrastructure resources
2. Determine and agree on which IT services and IT infrastructure resources are essential to sustain the operation of business processes. Analyse dependencies and identify weak links.   2.1 Determine which IT services and IT infrastructure resources are required to sustain the operation of key services and critical business processes.   2.2 Analyse dependencies and weak links across the 'full stack', i.e., from the top layer down to physical facilities.   2.3 Gain the consensus of business and IT leadership on the enterprise's most valued information and related technology assets.
3. Aggregate current risk scenarios by category, business line and functional area.   3.1 Inventory and evaluate IT process capability, skills and knowledge of people, and IT performance outcomes across the spectrum of IT risk (e.g., IT benefit/value enablement, IT programme and project delivery, IT operations and service delivery).   3.2 Determine where normal process execution can or cannot provide the right controls and the ability to take on acceptable risk.   3.3 Identify where reducing process outcome variability can contribute to a more robust internal control structure, improve IT and business performance, and exploit/seize opportunities.
4. On a regular basis, capture all risk profile information and consolidate it into an aggregated risk profile.   4.1 Review the collection of attributes and values across IT risk scenario components and their inherent connections to business impact categories.   4.2 Adjust entries based on changing risk conditions and emerging threats to IT benefit/value enablement, IT programme and project delivery, and IT operations and service delivery.   4.3 Update distributions and ranges based on asset/resource criticality, data on the operating environment and risk event data. Link event types to risk categories and business impact categories.   4.4 Aggregate event types by category, business line and functional area.   4.5 At a minimum, update the IT risk scenario components in response to any significant internal or external change, and review them annually.
5. Based on all risk profile data, define a set of risk indicators that allow the quick identification and monitoring of current risk and risk trends.   5.1 Capture the risk profile within tools such as an IT risk register and IT risk map.   5.2 Build out the risk profile via the results of enterprise IT risk assessment, risk scenario components, risk event data collection, ongoing risk analysis and independent IT assessment findings.   5.3 For individual IT risk register entries, update key attributes such as name, description, owner, expected/actual frequency and potential/actual magnitude of associated scenarios, potential/actual business impact, and disposition.
6. Capture information on IT risk events that have materialised, for inclusion in the IT risk profile of the enterprise.   6.1 Design metrics or indicators that can point to IT-related events and incidents that can significantly impact the business.   6.2 Base the indicators on a model of what compromises exposure and capability in risk management.   6.3 Provide management with an understanding of the useful and potentially key risk indicators.   6.4 Regularly review KRIs in use by management, and recommend adjustment for changing internal and external conditions.

AP012 Risk-specific Process Practices, Inputs/Outputs, Activities and Detailed Activities *(cont.)*				
	Risk-specific Inputs (in Addition to COBIT 5 Inputs)		**Risk-specific Outputs (in Addition to COBIT 5 Outputs)**	
Management Practice	**From**	**Description**	**Description**	**To**
AP012.04 Articulate risk. Provide information on the current state of IT-related exposures and opportunities in a timely manner to all required stakeholders for appropriate response.	Risk-specific inputs and outputs are not relevant for this practice. The generic COBIT 5 inputs and outputs can be used as further guidance.			

Risk-specific Activities (in Addition to COBIT 5 Activities)
1. Report the results of risk analysis to all affected stakeholders in terms and formats useful to support enterprise decisions. Wherever possible, include probabilities and ranges of loss or gain along with confidence levels that enable management to balance risk-return. 1.1 Co-ordinate additional risk analysis activity as required by decision makers (e.g., report rejection, scope adjustment). 1.2 Communicate clearly the risk-return context. 1.3 Identify the negative impacts of events/scenarios that should drive response decisions and the positive impacts of events/scenarios that represent opportunities management should channel back into the strategy and objective-setting process.
2. Provide decision makers with an understanding of worst-case and most probable scenarios, due diligence exposures, and significant reputation, legal or regulatory considerations. 2.1 In this effort, include the following: • Key components of risk (e.g., frequency, magnitude, impact) and key risk factors and their estimated effects • Estimated probable loss magnitude or probable future gain • Estimated high-end loss/gain potential and most probable loss/gain scenario(s) (e.g., a probable loss frequency of between three and five times per year, and a probable loss magnitude of between US $50,000 and $100,000, with 90 percent confidence) • Additional relevant information to support the conclusions and recommendations of the analysis
3. Report the current risk profile to all stakeholders, including effectiveness of the risk management process, control effectiveness, gaps, inconsistencies, redundancies, remediation status, and their impacts on the risk profile. 3.1 Meet the risk reporting needs of various stakeholders (e.g., board, risk committee, risk control functions, business unit management) by applying the principles of relevance, efficiency, timeliness and accuracy to the reports. 3.2 Include the following in the reporting: control effectiveness and performance, issues and gaps, remediation status, events and incidents and their impacts on the risk profile, performance of risk management processes. 3.3 Provide inputs to integrated enterprise reporting.
4. Review the results of objective third-party assessments, internal audit, and quality assurance reviews and map them to the risk profile. Review identified gaps and exposures to determine the need for additional risk analysis. 4.1 Take gaps and exposures to the business for their call on disposition or the need for risk analysis. 4.2 Help the business understand how corrective action plans will affect the overall risk profile. 4.3 Identify opportunities for integration with other remediation efforts and ongoing risk management activities.
5. On a periodic basis, for areas with relative risk and risk capacity parity, identify IT-related opportunities that would allow the acceptance of greater risk and enhanced growth and return. 5.1 Look for opportunities where IT can be used to: • Leverage enterprise resources in creating competitive advantage (e.g., use existing information in new ways, better leverage human and business resources). • Reduce enterprise co-ordination costs. • Exploit scale and scope economies in certain key strategic resources common to several lines of business. • Leverage structural differences with competitors. • Co-ordinate activities amongst business units or in the value chain.

	Risk-specific Inputs (in Addition to COBIT 5 Inputs)		Risk-specific Outputs (in Addition to COBIT 5 Outputs)	
AP012 Risk-specific Process Practices, Inputs/Outputs, Activities and Detailed Activities *(cont.)*				
Management Practice	**From**	**Description**	**Description**	**To**
AP012.05 Define a risk management action portfolio. Manage opportunities to reduce risk to an acceptable level as a portfolio.	Risk-specific inputs and outputs are not relevant for this practice. The generic COBIT 5 inputs and outputs can be used as further guidance.			

Risk-specific Activities (in Addition to COBIT 5 Activities)

1. Maintain an inventory of control activities that are in place to manage risk and that enable risk to be taken in line with risk appetite and tolerance. Classify control activities and map them to specific IT risk statements and aggregations of IT risk.
  1.1 Across the risk focus areas, inventory the controls in place to manage risk and enable risk to be taken in line with risk appetite and tolerance.
  1.2 Classify controls (e.g., predictive, preventive, detective, corrective) and map them to specific IT risk statements and aggregations of IT risk.

2. Determine if each organisational entity monitors risk and accepts accountability for operating within its individual and portfolio tolerance levels.
  2.1 Monitor operational alignment with risk tolerance thresholds.
  2.2 Ensure that each business line accepts accountability for operating within its individual and portfolio tolerance levels and for embedding monitoring tools into key operating processes.
  2.3 Monitor control performance, and measure variance from thresholds against objectives.

3. Define a balanced set of project proposals designed to reduce risk and/or projects that enable strategic enterprise opportunities, considering cost/benefits, effect on current risk profile and regulations.
  3.1 Respond to discovered risk exposure and opportunity.
  3.2 Select candidate IT controls based on specific threats, the degree of risk exposure, probable loss and mandatory requirements specified in IT standards.
  3.3 Monitor changes to the underlying business operational risk profiles and adjust the rankings of risk response projects.
  3.4 Communicate with key stakeholders early in the process.
  3.5 Conduct pilot testing and review performance data to verify operation against design.
  3.6 Map new and updated operational controls to monitoring mechanisms that will measure control performance over time, and prompt management corrective action when needed.
  3.7 Identify and train staff on new procedures as they are deployed.
  3.8 Report IT risk action plan progress. Monitor IT risk action plans at all levels to ensure the effectiveness of required actions and determine whether acceptance of residual risk was obtained.
  3.9 Ensure that committed actions are owned by the affected process owner(s) and deviations are reported to senior management.

	Risk-specific Inputs (in Addition to COBIT 5 Inputs)		Risk-specific Outputs (in Addition to COBIT 5 Outputs)	
Management Practice	**From**	**Description**	**Description**	**To**
AP012.06 Respond to risk. Respond in a timely manner with effective measures to limit the magnitude of loss from IT-related events.	Risk-specific inputs and outputs are not relevant for this practice. The generic COBIT 5 inputs and outputs can be used as further guidance.			

AP012 Risk-specific Process Practices, Inputs/Outputs, Activities and Detailed Activities *(cont.)*

Risk-specific Activities (in Addition to COBIT 5 Activities)

1. Prepare, maintain and test plans that document the specific steps to take when a risk event may cause a significant operational or development incident with serious business impact, including pathways of escalation across the enterprise.
   1.1 Prepare for the materialisation of threats through plans that document the specific steps to take when a risk event may cause an operational, developmental and/or strategic business impact (i.e., IT-related incident) or has already caused a business impact.
   1.2 Maintain open communication about risk acceptance, risk management activities, analysis techniques and results available to assist with plan preparation.
   1.3 When developing action plans, consider how long the enterprise may be exposed and how long it may take to recover.
   1.4 Define pathways of escalation across the enterprise, from line management to executive committees.
   1.5 Verify that incident response plans for highly critical processes are adequate.

2. Categorise incidents, and compare actual exposures against risk tolerance thresholds. Communicate business impacts to decision makers as part of reporting, and update the risk profile.
   2.1 Monitor IT risk. Monitor the environment.
   2.2 When a control limit has been breached, either escalate to the next step or confirm that the measure is back within limits.
   2.3 Ensure policy is followed and that there is clear accountability for follow-up actions.

3. Apply the appropriate response plan to minimise the impact when risk incidents occur.
   3.1 Initiate incident response.
   3.2 Take action to minimise the impact of an incident in progress.
   3.3 Identify the category of the incident and follow the steps in the response plan.
   3.4 Inform all stakeholders and affected parties that an incident is occurring.
   3.5 Identify the amount of time required to carry out the plan and make adjustments, as necessary, for the situation at hand.
   3.6 Ensure that the correct action is taken.

4. Examine past adverse events/losses and missed opportunities and determine root causes. Communicate root cause, additional risk response requirements and process improvements to risk governance processes and appropriate decision makers.
   4.1 Communicate lessons learned from risk events.
   4.2 Examine past adverse events/losses and missed opportunities.
   4.3 Determine whether there was a failure stemming from lack of awareness, capability or motivation.
   4.4 Research the root cause of similar risk events and the relative effectiveness of actions taken then and now.
   4.5 Determine the extent of any underlying problems.
   4.6 Identify tactical corrections; potential investments in projects; or adjustments to overall risk governance, evaluation and/or response processes.
   4.7 Integrate with the IT service desk and incident response process and the IT problem management process to identify and correct the underlying root cause.
   4.8 Identity the root cause of IT benefit/value enablement and IT programme and project delivery incidents through open communication across business and IT functions. Request additional risk analysis as needed.
   4.9 Communicate root cause, additional risk response requirements and process improvements to risk governance processes and appropriate decision makers.

APPENDIX D
USING COBIT 5 ENABLERS TO MITIGATE IT RISK SCENARIOS

Introduction

This appendix provides a number of examples on how COBIT 5 enablers can be used to respond to risk scenarios. The risk scenarios were identified in section 2B, chapter 3 of this publication.

In the risk response process, risk mitigation is identified as one of the options to respond to any excessive risk. IT risk mitigation is equivalent to implementing a number of IT controls. In COBIT 5 terms, IT controls can be any enabler, e.g., putting in place an organisational structure, putting in place certain governance or management practices or activities, etc.

For each of the risk categories, potential mitigating actions relating to all seven COBIT 5 enablers are provided, with a reference, title and description for each enabler that can help to mitigate the risk.

When using the examples in this appendix, the reader should keep in mind that:
• The tables do not replace the risk analysis exercise. The risk categories presented here are generic and in themselves can cover many derived and varying scenarios. Every enterprise first needs to customise and define its own set of risk scenarios.
• The tables need to be customised. Every situation is unique and every risk and all surrounding risk factors need to be considered before risk mitigation measures are defined.
• The suggested controls are not absolute. They need to be weighed in terms of cost/benefit, i.e., how effective they will be in reducing risk and what the cost is to implement. The effect of the mitigating action on potential impact and frequency of the risk should be estimated and depends on the maturity of the implementation, the context of the enterprise, etc. When effect on impact and frequency is estimated to be 'high', the action can be considered 'essential' for the enterprise.
• The suggested list of controls may not be complete for a particular situation, so the user should be prepared to carefully analyse whether any controls need to be added (or taken away) based on each situation. For some scenarios, additional and more detailed guidance may be required. Examples are information security risk items and controls such as vulnerability management or application security scanning.

The value of this section ties into:
• **Risk assessment and analysis**—When frequency and impact need to be assessed, controls/enablers need to be taken into account to determine the impact and a realistic frequency assessment. Weak or strong enablers are very important risk factors.
• **Risk mitigation**—When risk requires mitigation, i.e., controls/enablers practices need to be defined and implemented. The tables provide a number of suggested controls that can help mitigate the risk at hand.

Note: The tables linking each risk category to a set of mitigating enablers stays at a very generic level, thus providing a starting point for analysing and mitigating risk. Each enterprise will need to tailor the set of enablers required to analyse and mitigate each specific risk scenario they have to handle.

D.1. Scenario 1: Portfolio Establishment and Maintenance		
Risk Scenario Category	Portfolio establishment and maintenance	
Principles, Policies and Frameworks Enabler		
Reference	**Contribution to Response to Scenario**	
Programme/project management policy	To enforce the use of the overall programme/project methodology including corporate policy on business case or due diligence in order to improve the visibility of the relative value of programmes (compared to each other). This policy should describe approval investment thresholds for programme value.	
Process Enabler		
Reference	**Title**	**Management Practice**
EDM02.01	Evaluate value optimisation.	Continually evaluate the portfolio of IT-enabled investments, services and assets to determine the likelihood of achieving enterprise objectives and delivering value at a reasonable cost. Identify and make judgement on any changes in direction that need to be given to management to optimise value creation.
EDM02.02	Direct value optimisation.	Direct value management principles and practices to enable optimal value realisation from IT-enabled investments throughout their full economic life cycle.
EDM02.03	Monitor value optimisation.	Monitor the key goals and metrics to determine the extent to which the business is generating the expected value and benefits to the enterprise from IT-enabled investments and services. Identify significant issues and consider corrective actions.
APO01.01	Define the organisational structure.	Establish an internal and extended organisational structure that reflects business needs and IT priorities. Put in place the required management structures (e.g., committees) that enable management decision making to take place in the most effective and efficient manner.
APO01.04	Communicate management objectives and direction.	Communicate awareness and understanding of IT objectives and direction to appropriate stakeholders and users throughout the enterprise.
APO02.03	Define the target IT capabilities.	Define the target business and IT capabilities and required IT services. This should be based on the understanding of the enterprise environment and requirements; the assessment of the current business process and IT environment and issues; and consideration of reference standards, best practices and validated emerging technologies or innovation proposals.
APO04.03	Monitor and scan the technology environment.	Perform systematic monitoring and scanning of the enterprise's external environment to identify emerging technologies that have the potential to create value (e.g., by realising the enterprise strategy, optimising costs, avoiding obsolescence, and better enabling enterprise and IT processes). Monitor the marketplace, competitive landscape, industry sectors, and legal and regulatory trends to be able to analyse emerging technologies or innovation ideas in the enterprise context.
APO05.01	Establish the target investment mix.	Review and ensure clarity of the enterprise and IT strategies and current services. Define an appropriate investment mix based on cost, alignment with strategy, and financial measures such as cost and expected ROI over the full economic life cycle, degree of risk, and type of benefit for the programmes in the portfolio. Adjust the enterprise and IT strategies where necessary.
APO05.03	Evaluate and select programmes to fund.	Based on the overall investment portfolio mix requirements, evaluate and prioritise programme business cases, and decide on investment proposals. Allocate funds and initiate programmes.
APO05.05	Maintain portfolios.	Maintain portfolios of investment programmes and projects, IT services and IT assets.
APO06.02	Prioritise resource allocation.	Implement a decision-making process to prioritise the allocation of resources and rules for discretionary investments by individual business units. Include the potential use of external service providers and consider the buy, develop and rent options.
BAI02.01	Define and maintain business functional and technical requirements.	Based on the business case, identify, prioritise, specify and agree on business information, functional, technical and control requirements covering the scope/understanding of all initiatives required to achieve the expected outcomes of the proposed IT-enabled business solution.
Organisational Structures Enabler		
Reference	**Contribution to Response to Scenario**	
Programme and project management office (PMO)	Responsible for the quality of the business cases	
Board	Approval is required when programmes surpass a certain value threshold and risk level.	
CFO	Help with alignment of strategy and priorities, overall view on programmes.	

D.1. Scenario 1: Portfolio Establishment and Maintenance *(cont.)*	
Culture, Ethics and Behaviour Enabler	
Reference	**Contribution to Response to Scenario**
Programme selection includes data-driven decisions	Emotion and politics will not be a dominant factor in the decision making.
Stakeholder engagement	The full range of success factors will be taken into account when selecting programmes.
Focus on enterprise objectives	Ensure alignment with corporate strategy and priorities.
Information Enabler	
Reference	**Contribution to Response to Scenario**
Programme business case	Improves the visibility of the relative value of programmes (compared to each other)
Defined investment mix	Improves the visibility of the relative value of programmes (compared to each other)
Services, Infrastructure and Applications Enabler	
Reference	**Contribution to Response to Scenario**
Portfolio management tools	Decrease complexity and increase overview on programmes and projects.
People, Skills and Competencies Enabler	
Reference	**Contribution to Response to Scenario**
Programme/project finance skills	Create visibility on programme value.
Business requirements analysis	Transparency on enterprise strategy, related business requirements and priorities
Marketing-related skills	Create visibility on programme value.

D.2. Scenario 2: Programme/Project Life Cycle Management		
Risk Scenario Category	Programme/project life cycle management (programme/project initiation, economics, delivery, quality and termination)	
Principles, Policies and Frameworks Enabler		
Reference	**Contribution to Response to Scenario**	
Programme/project management policy	Measuring visibility and true status for decision makers should be based on common language and methodology: • Awareness regarding failing projects (in terms of cost, delays, scope creep, changed business priorities, etc.) and create information flows to induce corrective action. • To prevent fails scope changes to existing projects need to be managed strictly	

Process Enabler		
Reference	**Title**	**Management Practice**
EDM02.03	Monitor value optimisation.	Monitor the key goals and metrics to determine the extent to which the business is generating the expected value and benefits to the enterprise from IT-enabled investments and services. Identify significant issues and consider corrective actions.
APO01.01	Define the organisational structure.	Establish an internal and extended organisational structure that reflects business needs and IT priorities. Put in place the required management structures (e.g., committees) that enable management decision making to take place in the most effective and efficient manner.
APO06.04	Model and allocate costs.	Establish and use an IT costing model based on the service definition, ensuring that allocation of costs for services is identifiable, measurable and predictable, to encourage the responsible use of resources including those provided by service providers. Regularly review and benchmark the appropriateness of the cost/chargeback model to maintain its relevance and appropriateness to the evolving business and IT activities.
APO06.05	Manage costs.	Implement a cost management process comparing actual costs to budgets. Costs should be monitored and reported and, in the case of deviations, identified in a timely manner and their impact on enterprise processes and services assessed.
BAI01.01	Maintain a standard approach for programme and project management.	Maintain a standard approach for programme and project management that enables governance and management review and decision making and delivery management activities focused on achieving value and goals (requirements, risk, costs, schedule, quality) for the business in a consistent manner.
BAI01.02	Initiate a programme.	Initiate a programme to confirm the expected benefits and obtain authorisation to proceed. This includes agreeing on programme sponsorship, confirming the programme mandate through approval of the conceptual business case, appointing programme board or committee members, producing the programme brief, reviewing and updating the business case, developing a benefits realisation plan, and obtaining approval from sponsors to proceed.
BAI01.03	Manage stakeholder engagement.	Manage stakeholder engagement to ensure an active exchange of accurate, consistent and timely information that reaches all relevant stakeholders. This includes planning, identifying and engaging stakeholders and managing their expectations.
BAI01.04	Develop and maintain the programme plan.	Formulate a programme to lay the initial groundwork and to position it for successful execution by formalising the scope of the work to be accomplished and identifying the deliverables that will satisfy its goals and deliver value. Maintain and update the programme plan and business case throughout the full economic life cycle of the programme, ensuring alignment with strategic objectives and reflecting the current status and updated insights gained to date.
BAI01.05	Launch and execute the programme.	Launch and execute the programme to acquire and direct the resources needed to accomplish the goals and benefits of the programme as defined in the programme plan. In accordance with stage-gate or release review criteria, prepare for stage-gate, iteration or release reviews to report on the progress of the programme and to be able to make the case for funding up to the following stage-gate or release review.
BAI01.06	Monitor, control and report on the programme outcomes.	Monitor and control programme (solution delivery) and enterprise (value/outcome) performance against plan throughout the full economic life cycle of the investment. Report this performance to the programme steering committee and the sponsors.
BAI01.07	Start up and initiate projects within a programme.	Define and document the nature and scope of the project to confirm and develop amongst stakeholders a common understanding of project scope and how it relates to other projects within the overall IT-enabled investment programme. The definition should be formally approved by the programme and project sponsors.
BAI01.08	Plan projects.	Establish and maintain a formal, approved integrated project plan (covering business and IT resources) to guide project execution and control throughout the life of the project. The scope of projects should be clearly defined and tied to building or enhancing business capability.
BAI01.09	Manage programme and project quality.	Prepare and execute a quality management plan, processes and practices, aligned with the QMS that describes the programme and project quality approach and how it will be implemented. The plan should be formally reviewed and agreed on by all parties concerned and then incorporated into the integrated programme and project plans.

D.2. Scenario 2: Programme/Project Life Cycle Management *(cont.)*		
Process Enabler *(cont.)*		
Reference	Title	Management Practice
BAI01.10	Manage programme and project risk.	Eliminate or minimise specific risk associated with programmes and projects through a systematic process of planning, identifying, analysing, responding to and monitoring and controlling the areas or events that have the potential to cause unwanted change. Risk faced by programme and project management should be established and centrally recorded.
BAI01.11	Monitor and control projects.	Measure project performance against key project performance criteria such as schedule, quality, cost and risk. Identify any deviations from the expected. Assess the impact of deviations on the project and overall programme, and report results to key stakeholders.
BAI01.12	Manage project resources and work packages.	Manage project work packages by placing formal requirements on authorising and accepting work packages, and assigning and co-ordinating appropriate business and IT resources.
BAI01.13	Close a project or iteration.	At the end of each project, release or iteration, require the project stakeholders to ascertain whether the project, release or iteration delivered the planned results and value. Identify and communicate any outstanding activities required to achieve the planned results of the project and the benefits of the programme, and identify and document lessons learned for use on future projects, releases, iterations and programmes.

Organisational Structures Enabler	
Reference	Contribution to Response to Scenario
Programme and project management office (PMO)	Ensure consistency of approach within programme/project monitoring.
CIO	Take corrective action if required.
Programme/project sponsor	Overall accountable for budget tracking and value demonstration
Programme/project manager	Overall responsible for budget tracking and value demonstration

Culture, Ethics and Behaviour Enabler	
Reference	Contribution to Response to Scenario
Programme/project monitoring includes data-driven activities	Emotions and politics will not be a dominant factor in the decision making
Admitting to bad news is supported by senior management	Enables earlier decision making and minimises impact
Programme benefit realisation plan	This input will provide the necessary data to track the progress and estimate potential overrun.
Programme budget and benefits register	This input will provide the necessary data to track the progress and estimate potential overrun.
Programme status report	Measuring visibility and true status for decision makers should be based on common language and methodology.

Services, Infrastructure and Applications Enabler	
Reference	Contribution to Response to Scenario
Portfolio management tools	Increase transparency on budgetary status.

People, Skills and Competencies Enabler	
Reference	Contribution to Response to Scenario
Performance to budget control skills	The correct analytical skills will allow estimation of the consequences of failing projects such as potential budget overruns.

D.3. Scenario 3: IT Investment Decision Making	
Risk Scenario Category	IT investment decision making

Principles, Policies and Frameworks Enabler	
Reference	**Contribution to Response to Scenario**
Programme/Project management policy	The policy should define who needs to be involved in investment decisions and the chain of approval.

Process Enabler		
Reference	**Title**	**Management Practice**
APO05.06	Manage benefits achievement.	Monitor the benefits of providing and maintaining appropriate IT services and capabilities, based on the agreed-on and current business case.
APO06.02	Prioritise resource allocation.	Implement a decision-making process to prioritise the allocation of resources and rules for discretionary investments by individual business units. Include the potential use of external service providers and consider the buy, develop and rent options.
APO06.03	Create and maintain budgets.	Prepare a budget reflecting the investment priorities supporting strategic objectives based on the portfolio of IT-enabled programmes and IT services.
APO07.01	Maintain adequate and appropriate staffing.	Evaluate staffing requirements on a regular basis or on major changes to the enterprise or operational or IT environments to ensure that the enterprise has sufficient human resources to support enterprise goals and objectives. Staffing includes both internal and external resources.
BAI01.03	Manage stakeholder engagement.	Manage stakeholder engagement to ensure an active exchange of accurate, consistent and timely information that reaches all relevant stakeholders. This includes planning, identifying and engaging stakeholders and managing their expectations.
BAI03.04	Procure solution components.	Procure solution components based on the acquisition plan in accordance with requirements and detailed designs, architecture principles and standards, and the enterprise's overall procurement and contract procedures, QA requirements, and approval standards. Ensure that all legal and contractual requirements are identified and addressed by the supplier.

Organisational Structures Enabler	
Reference	**Contribution to Response to Scenario**
CIO	Accountable for proper investment decision making
CFO	Accountable for proper investment decision making

Culture, Ethics and Behaviour Enabler	
Reference	**Contribution to Response to Scenario**
Decision-making process is data driven	Emotions and politics will not be a dominant factor in the decision making.

Information Enabler	
Reference	**Contribution to Response to Scenario**
Business cases	Clarify the purpose, cost and return on investment of IT initiatives.
Prioritisation and ranking of IT initiatives	Overview of IT initiatives to facilitate selection
IT budget and plan	Overview on available IT budget and guidelines
N/A	N/A

People, Skills and Competencies Enabler	
Reference	**Contribution to Response to Scenario**
Cost allocation and budgeting	Ability to detail financial aspects of IT initiatives
Business case analysis	Clarify the purpose, cost and return on investment of IT initiatives.

D.4. Scenario 4: IT Expertise and Skills		
Risk Scenario Category	IT expertise and skills	
Principles, Policies and Frameworks Enabler		
Reference	**Contribution to Response to Scenario**	
HR policy	Describes the requirements development for selecting and evaluating IT profiles throughout the entire career.	
Process Enabler		
Reference	**Title**	**Management Practice**
APO01.01	Define the organisational structure.	Establish an internal and extended organisational structure that reflects business needs and IT priorities. Put in place the required management structures (e.g., committees) that enable management decision making to take place in the most effective and efficient manner.
APO01.04	Communicate management objectives and direction.	Communicate awareness and understanding of IT objectives and direction to appropriate stakeholders and users throughout the enterprise.
APO02.01	Understand enterprise direction.	Consider the current enterprise environment and business processes, as well as the enterprise strategy and future objectives. Consider also the external environment of the enterprise (industry drivers, relevant regulations, basis for competition).
APO03.01	Develop the enterprise architecture vision.	The architecture vision provides a first-cut, high-level description of the baseline and target architectures, covering the business, information, data, application and technology domains. The architecture vision provides the sponsor with a key tool to sell the benefits of the proposed capability to stakeholders within the enterprise. The architecture vision describes how the new capability will meet enterprise goals and strategic objectives and address stakeholder concerns when implemented.
APO07.01	Maintain adequate and appropriate staffing.	Evaluate staffing requirements on a regular basis or on major changes to the enterprise or operational or IT environments to ensure that the enterprise has sufficient human resources to support enterprise goals and objectives. Staffing includes both internal and external resources.
APO07.02	Identify key IT personnel.	Identify key IT personnel while minimising reliance on a single individual performing a critical job function through knowledge capture (documentation), knowledge sharing, succession planning and staff backup.
APO07.03	Maintain the skills and competencies of personnel.	Define and manage the skills and competencies required of personnel. Regularly verify that personnel have the competencies to fulfil their roles on the basis of their education, training and/or experience, and verify that these competencies are being maintained, using qualification and certification programmes where appropriate. Provide employees with ongoing learning and opportunities to maintain their knowledge, skills and competencies at a level required to achieve enterprise goals.
APO07.04	Evaluate employee job performance.	Perform timely performance evaluations on a regular basis against individual objectives derived from the enterprise's goals, established standards, specific job responsibilities, and the skills and competency framework. Employees should receive coaching on performance and conduct whenever appropriate.
APO07.05	Plan and track the usage of IT and business human resources.	Understand and track the current and future demand for business and IT human resources with responsibilities for enterprise IT. Identify shortfalls and provide input into sourcing plans, enterprise and IT recruitment processes sourcing plans, and business and IT recruitment processes.
Organisational Structures Enabler		
Reference	**Contribution to Response to Scenario**	
CIO	Responsible for gap analysis regarding IT skills and competencies	
Head of HR	Responsible for establishing expectations towards staff	
Specific IT management functions	Responsible for identifying specific requirements	
Culture, Ethics and Behaviour Enabler		
Reference	**Contribution to Response to Scenario**	
Awareness of business activities by IT staff	IT staff should know the core business activities of the enterprise they support.	
Foster competency development with IT staff	Continuous development of existing IT skills.	

D.4. Scenario 4: IT Expertise and Skills *(cont.)*	
Information Enabler	
Reference	**Contribution to Response to Scenario**
Skills and competencies matrix	Describe the existing skills and competencies within the IT organisation and allow for gap analysis
Competency and career/skills development plans	Describe the required evolution of specific IT profiles.
Generic function descriptions	Describe skills/experience and knowledge requirements for generic profiles within the IT organisations.
Knowledge repositories	Minimizing the effect of partial unavailability of resources by sharing knowledge regarding processes, technology, etc.
Services, Infrastructure and Applications Enabler	
Reference	**Contribution to Response to Scenario**
N/A	N/A
People, Skills and Competencies Enabler	
Reference	**Contribution to Response to Scenario**
HR skills	Management of skills and competencies
Business analysis	Matching the business needs to the required IT skills

D.5. Scenario 5: Staff Operations	
Risk Scenario Category	Staff operations (human error and malicious intent)
Organisational Structures Enabler	
Reference	**Contribution to Response to Scenario**
Information security manager	Responsible for technical protection of assets and information
Head of HR	Responsible for establishing expectations towards staff
Head of IT operations	Responsible for managing the operational environment
Culture, Ethics and Behaviour Enabler	
Reference	**Contribution to Response to Scenario**
Everybody is responsible for the protection of information within the enterprise	Leading by example
People respect the importance of policies and procedures	Preventing errors and accidents
Information Enabler	
Reference	**Contribution to Response to Scenario**
Staffing contract	Contractual obligations, restrictions and rights of the staff
Access and event logs	Detect wrongful activity.
Allocated roles and responsibilities/levels of authority	Provide clarity on organisational distribution.
Services, Infrastructure and Applications Enabler	
Reference	**Contribution to Response to Scenario**
Access control	To prevent unauthorised physical access
Alarm and monitoring security system	To prevent unauthorised physical access
People, Skills and Competencies Enabler	
Reference	**Contribution to Response to Scenario**
Security skills	Prevent malicious intent.

D.6. Scenario 6: Information		
Risk Scenario Category	Information (damage, leakage and access)	
Principles, Policies and Frameworks Enabler		
Reference	**Contribution to Response to Scenario**	
Physical security policy	Access can only be provided to authorised staff.	
Backup policy	Backups are available.	
Business continuity and disaster recovery policy	Validate recoverability of data.	
Information security policy	Defines technical limitations on sharing and using information.	
Process Enabler		
Reference	**Title**	**Management Practice**
APO01.06	Define information (data) and system ownership.	Define and maintain responsibilities for ownership of information (data) and information systems. Ensure that owners make decisions about classifying information and systems and protecting them in line with this classification.
BAI02.01	Define and maintain business functional and technical requirements	Based on the business case, identify, prioritise, specify and agree on business information, functional, technical and control requirements covering the scope/understanding of all initiatives required to achieve the expected outcomes of the proposed IT-enabled business solution.
BAI04.05	Investigate and address availability, performance and capacity issues.	Address deviations by investigating and resolving identified availability, performance and capacity issues.
DSS01.01	Perform operational procedures.	Maintain and perform operational procedures and operational tasks reliably and consistently.
DSS01.05	Manage facilities.	Manage facilities, including power and communications equipment, in line with laws and regulations, technical and business requirements, vendor specifications, and health and safety guidelines.
DSS04.03	Develop and implement a business continuity response.	Develop a business continuity plan (BCP) based on the strategy that documents the procedures and information in readiness for use in an incident to enable the enterprise to continue its critical activities.
DSS04.04	Exercise, test and review the BCP.	Test the continuity arrangements on a regular basis to exercise the recovery plans against predetermined outcomes and to allow innovative solutions to be developed and help to verify over time that the plan will work as anticipated.
DSS05.02	Manage network and connectivity security.	Use security measures and related management procedures to protect information over all methods of connectivity.
DSS05.05	Manage physical access to IT assets.	Define and implement procedures to grant, limit and revoke access to premises, buildings and areas according to business needs, including emergencies. Access to premises, buildings and areas should be justified, authorised, logged and monitored. This should apply to all persons entering the premises, including staff, temporary staff, clients, vendors, visitors or any other third party.
DSS05.06	Manage sensitive documents and output devices.	Establish appropriate physical safeguards, accounting practices and inventory management over sensitive IT assets, such as special forms, negotiable instruments, special-purpose printers or security tokens.
DSS06.04	Manage errors and exceptions.	Manage business process exceptions and errors and facilitate their correction. Include escalation of business process errors and exceptions and the execution of defined corrective actions. This provides assurance of the accuracy and integrity of the business information process.
DSS06.05	Ensure traceability of Information events and accountabilities.	Ensure that business information can be traced to the originating business event and accountable parties. This enables traceability of the information through its life cycle and related processes. This provides assurance that information that drives the business is reliable and has been processed in accordance with defined objectives.
Organisational Structures Enabler		
Reference	**Contribution to Response to Scenario**	
Information security manager	Provide with guidance on proper controls and measures to protect data and hardware.	
Head of IT operations	Responsible for implementing proper controls and measures to protect data and hardware	

D.6. Scenario 6: Information *(cont.)*	
Culture, Ethics and Behaviour Enabler	
Reference	**Contribution to Response to Scenario**
Information security is practiced in daily operations	Always select the safest option with regard to daily operations.
Need to access only	Limit the access of staff without affecting performance.
Everybody is responsible for the protection of information within the enterprise	Lead by example.
Information Enabler	
Reference	**Contribution to Response to Scenario**
Backup reports	Describes the status regarding backups.
Data loss prevention campaigns	Increase awareness within the enterprise.
Non-disclosure agreements	Contractually protect IP by deterring staff from disclosing information to malicious parties.
Access and event logs	Detect wrongful activity.
Services, Infrastructure and Applications Enabler	
Reference	**Contribution to Response to Scenario**
Access control	To prevent unauthorised physical access
Backup systems	Ensure proper recovery in case of loss, modification or corruption of data.
Data protection infrastructure and applications	Encryption, passwords, email monitoring, etc., to apply the need-to-know principle
People, Skills and Competencies Enabler	
Reference	**Contribution to Response to Scenario**
Technical skills	Regarding the proper controls and measures to protect data and hardware (e.g., data backup, storage)
Information security policy	Defines technical limitations on sharing and using information.

D.7. Scenario 7: Architecture	
Risk Scenario Category	Architecture (architectural vision and design)

Principles, Policies and Frameworks Enabler	
Reference	**Contribution to Response to Scenario**
Architecture principles	Architecture principles define the underlying general rules and guidelines for the use and deployment of all IT resources and assets across the enterprise.
Exceptions procedure	In specific cases exceptions to the existing architectural rules can be allowed. Specific cases and the procedure to follow for approval should be described.

Process Enabler		
Reference	**Title**	**Management Practice**
APO02.01	Understand enterprise direction.	Consider the current enterprise environment and business processes, as well as the enterprise strategy and future objectives. Consider also the external environment of the enterprise (industry drivers, relevant regulations, basis for competition).
APO02.03	Define the target IT capabilities.	Define the target business and IT capabilities and required IT services. This should be based on the understanding of the enterprise environment and requirements; the assessment of the current business process and IT environment and issues; and consideration of reference standards, best practices and validated emerging technologies or innovation proposals.
APO03.01	Develop the enterprise architecture vision.	The architecture vision provides a first-cut, high-level description of the baseline and target architectures, covering the business, information, data, application and technology domains. The architecture vision provides the sponsor with a key tool to sell the benefits of the proposed capability to stakeholders within the enterprise. The architecture vision describes how the new capability will meet enterprise goals and strategic objectives and address stakeholder concerns when implemented.
APO03.02	Define reference architecture.	The reference architecture describes the current and target architectures for the business, information, data, application and technology domains.
APO03.03	Select opportunities and solutions.	Rationalise the gaps between baseline and target architectures, taking both business and technical perspectives, and logically group them into project work packages. Integrate the project with any related IT-enabled investment programmes to ensure that the architectural initiatives are aligned with and enable these initiatives as part of overall enterprise change. Make this a collaborative effort with key enterprise stakeholders from business and IT to assess the enterprise's transformation readiness, and identify opportunities, solutions and all implementation constraints.
APO03.04	Define architecture implementation.	Create a viable implementation and migration plan in alignment with the programme and project portfolios. Ensure that the plan is closely co-ordinated to ensure that value is delivered and the required resources are available to complete the necessary work.
APO03.05	Provide enterprise architecture services.	The provision of enterprise architecture services within the enterprise includes guidance to and monitoring of implementation projects, formalising ways of working through architecture contracts, and measuring and communicating architecture's value-add and compliance monitoring.
APO04.03	Monitor and scan the technology environment.	Perform systematic monitoring and scanning of the enterprise's external environment to identify emerging technologies that have the potential to create value (e.g., by realising the enterprise strategy, optimising costs, avoiding obsolescence, and better enabling enterprise and IT processes). Monitor the marketplace, competitive landscape, industry sectors, and legal and regulatory trends to be able to analyse emerging technologies or innovation ideas in the enterprise context.
APO04.04	Assess the potential of emerging technologies and innovation ideas.	Analyse identified emerging technologies and/or other IT innovation suggestions. Work with stakeholders to validate assumptions on the potential of new technologies and innovation.
APO04.06	Monitor the implementation and use of innovation.	Monitor the implementation and use of emerging technologies and innovations during integration, adoption and for the full economic life cycle to ensure that the promised benefits are realised and to identify lessons learned.

Organisational Structures Enabler	
Reference	**Contribution to Response to Scenario**
Architecture board	Ensure compliance with the target architecture and allow exceptions when needed.

Culture, Ethics and Behaviour Enabler	
Reference	**Contribution to Response to Scenario**
Respect agreed-on standards	The enterprise should stimulate the use of agreed-on standards.

Information Enabler	
Reference	**Contribution to Response to Scenario**
Architecture model	Target architecture model

D.7. Scenario 7: Architecture *(cont.)*	
Services, Infrastructure and Applications Enabler	
Reference	**Contribution** to Response to Scenario
CMDB	Configuration management database
Architecture modelling software	Modelling application will optimize the architecture development and minimize the effort of analysing impact to architecture in case of exceptions or changes.
People, Skills and Competencies Enabler	
Reference	**Contribution** to Response to Scenario
Leadership and communication	Clarify the rationale for the architecture and the potential consequences.
Architecture skills	Develop efficient and effective architecture aligned to the business requirements.

D.8. Scenario 8: Infrastructure	

Risk Scenario Category	Infrastructure (hardware, operating system and controlling technology) (selection/implementation, operations and decommissioning)

Principles, Policies and Frameworks Enabler

Reference	Contribution to Response to Scenario
Architecture principles	Define the underlying general rules and guidelines for the use and deployment of all IT resources and assets across the enterprise.
Change management policy	Guide changes and evolutions in infrastructure.

Process Enabler

Reference	Title	Management Practice
APO02.03	Define the target IT capabilities.	Define the target business and IT capabilities and required IT services. This should be based on the understanding of the enterprise environment and requirements; the assessment of the current business process and IT environment and issues; and consideration of reference standards, best practices and validated emerging technologies or innovation proposals.
APO04.03	Monitor and scan the technology environment.	Perform systematic monitoring and scanning of the enterprise's external environment to identify emerging technologies that have the potential to create value (e.g., by realising the enterprise strategy, optimising costs, avoiding obsolescence, and better enabling enterprise and IT processes). Monitor the marketplace, competitive landscape, industry sectors, and legal and regulatory trends to be able to analyse emerging technologies or innovation ideas in the enterprise context.
BAI03.03	Develop solution components.	Develop solution components progressively in accordance with detailed designs following development methods and documentation standards, quality assurance (QA) requirements, and approval standards. Ensure that all control requirements in the business processes, supporting IT applications and infrastructure services, services and technology products, and partners/suppliers are addressed.
BAI04.01	Assess current availability, performance and capacity and create a baseline.	Assess availability, performance and capacity of services and resources to ensure that cost-justifiable capacity and performance are available to support business needs and deliver against SLAs. Create availability, performance and capacity baselines for future comparison.
BAI04.02	Assess business impact.	Identify important services to the enterprise, map services and resources to business processes, and identify business dependencies. Ensure that the impact of unavailable resources is fully agreed on and accepted by the customer. Ensure that, for vital business functions, the SLA availability requirements can be satisfied.
BAI04.03	Plan for new or changed service requirements.	Plan and prioritise availability, performance and capacity implications of changing business needs and service requirements.
BAI04.04	Monitor and review availability and capacity.	Monitor, measure, analyse, report and review availability, performance and capacity. Identify deviations from established baselines. Review trend analysis reports identifying any significant issues and variances, initiating actions where necessary, and ensuring that all outstanding issues are followed up.
BAI04.05	Investigate and address availability, performance and capacity issues.	Address deviations by investigating and resolving identified availability, performance and capacity issues.
BAI10.04	Produce status and configuration reports.	Define and produce configuration reports on status changes of configuration items.
BAI10.05	Verify and review integrity of the configuration repository.	Periodically review the configuration repository and verify completeness and correctness against the desired target.
DSS05.05	Manage physical access to IT assets.	Define and implement procedures to grant, limit and revoke access to premises, buildings and areas according to business needs, including emergencies. Access to premises, buildings and areas should be justified, authorised, logged and monitored. This should apply to all persons entering the premises, including staff, temporary staff, clients, vendors, visitors or any other third party.

Organisational Structures Enabler

Reference	Contribution to Response to Scenario
Head of IT operations	Accountable for the proper management and maintenance of the IT infrastructure
Head of architecture	Designing architecture in an optimal way

Culture, Ethics and Behaviour Enabler

Reference	Contribution to Response to Scenario
Respect the available assets	All staff is required to maintain the assets in an appropriate manner

D.8. Scenario 8: Infrastructure *(cont.)*	
Information Enabler	
Reference	**Contribution to Response to Scenario**
Architecture model	Target architecture model
(Updates to) asset inventory	Tracking all assets throughout the enterprise
Maintenance plan	Planning the maintenance of the IT infrastructure
Configuration status reports	Tracking changes to configuration
Services, Infrastructure and Applications Enabler	
Reference	**Contribution to Response to Scenario**
CMDB	Assists in identifying areas for improvement.
People, Skills and Competencies Enabler	
Reference	**Contribution to Response to Scenario**
Architecture skills	Develop efficient and effective architecture aligned to the business requirements.
Technical skills	Managing the different infrastructure components

D.9. Scenario 9: Software	
Risk Scenario Category	Software (selection/implementation, operations and decommissioning)

Principles, Policies and Frameworks Enabler	
Reference	**Contribution to Response to Scenario**
Change management policy	Guiding changes and evolutions in infrastructure
Fallback procedure	Guidelines in case of roll back
Architecture principles	Architecture principles define the underlying general rules and guidelines for the use and deployment of all IT resources and assets across the enterprise.

Process Enabler		
Reference	**Title**	**Management Practice**
BAI03.01	Design high-level solutions.	Develop and document high-level designs using agreed-on and appropriate phased or rapid agile development techniques. Ensure alignment with the IT strategy and enterprise architecture. Reassess and update the designs when significant issues occur during detailed design or building phases or as the solution evolves. Ensure that stakeholders actively participate in the design and approve each version.
BAI03.02	Design detailed solution components.	Develop, document and elaborate detailed designs progressively using agreed-on and appropriate phased or rapid agile development techniques, addressing all components (business processes and related automated and manual controls, supporting IT applications, infrastructure services and technology products, and partners/suppliers). Ensure that the detailed design includes internal and external SLAs and OLAs.
BAI03.03	Develop solution components.	Develop solution components progressively in accordance with detailed designs following development methods and documentation standards, quality assurance (QA) requirements, and approval standards. Ensure that all control requirements in the business processes, supporting IT applications and infrastructure services, services and technology products, and partners/suppliers are addressed.
BAI03.05	Build solutions.	Install and configure solutions and integrate with business process activities. Implement control, security and auditability measures during configuration, and during integration of hardware and infrastructural software, to protect resources and ensure availability and data integrity. Update the services catalogue to reflect the new solutions.
BAI03.06	Perform quality assurance (QA).	Develop, resource and execute a QA plan aligned with the QMS to obtain the quality specified in the requirements definition and the enterprise's quality policies and procedures.
BAI03.07	Prepare for solution testing.	Establish a test plan and required environments to test the individual and integrated solution components, including the business processes and supporting services, applications and infrastructure.
BAI03.08	Execute solution testing.	Execute testing continually during development, including control testing, in accordance with the defined test plan and development practices in the appropriate environment. Engage business process owners and end users in the test team. Identify, log and prioritise errors and issues identified during testing.
BAI03.09	Manage changes to requirements.	Track the status of individual requirements (including all rejected requirements) throughout the project life cycle and manage the approval of changes to requirements.
BAI03.10	Maintain solutions.	Develop and execute a plan for the maintenance of solution and infrastructure components. Include periodic reviews against business needs and operational requirements.
BAI05.05	Enable operation and use.	Plan and implement all technical, operational and usage aspects such that all those who are involved in the future state environment can exercise their responsibility.
BAI06.01	Evaluate, prioritise and authorise change requests.	Evaluate all requests for change to determine the impact on business processes and IT services, and to assess whether change will adversely affect the operational environment and introduce unacceptable risk. Ensure that changes are logged, prioritised, categorised, assessed, authorised, planned and scheduled.
BAI06.02	Manage emergency changes.	Carefully manage emergency changes to minimise further incidents and make sure the change is controlled and takes place securely. Verify that emergency changes are appropriately assessed and authorised after the change.
BAI06.03	Track and report change status.	Maintain a tracking and reporting system to document rejected changes, communicate the status of approved and in-process changes, and complete changes. Make certain that approved changes are implemented as planned.
BAI06.04	Close and document the changes.	Whenever changes are implemented, update accordingly the solution and user documentation and the procedures affected by the change.
BAI07.01	Establish an implementation plan.	Establish an implementation plan that covers system and data conversion, acceptance testing criteria, communication, training, release preparation, promotion to production, early production support, a fallback/backout plan, and a post-implementation review. Obtain approval from relevant parties.
BAI07.03	Plan acceptance tests.	Establish a test plan based on enterprisewide standards that define roles, responsibilities, and entry and exit criteria. Ensure that the plan is approved by relevant parties.

D.9. Scenario 9: Software *(cont.)*		
Process Enabler *(cont.)*		
Reference	**Title**	**Management Practice**
BAI07.05	Perform acceptance tests.	Test changes independently in accordance with the defined test plan prior to migration to the live operational environment.
BAI07.08	Perform a post-implementation review.	Conduct a post-implementation review to confirm outcome and results, identify lessons learned, and develop an action plan. Evaluate and check the actual performance and outcomes of the new or changed service against the predicted performance and outcomes (i.e., the service expected by the user or customer).
BAI08.01	Nurture and facilitate a knowledge-sharing culture.	Devise and implement a scheme to nurture and facilitate a knowledge-sharing culture.
BAI08.04	Use and share knowledge.	Propagate available knowledge resources to relevant stakeholders and communicate how these resources can be used to address different needs (e.g., problem solving, learning, strategic planning and decision making).
BAI10.04	Produce status and configuration reports.	Define and produce configuration reports on status changes of configuration items.
BAI10.05	Verify and review integrity of the configuration repository.	Periodically review the configuration repository and verify completeness and correctness against the desired target.
Organisational Structures Enabler		
Reference	**Contribution to Response to Scenario**	
Head of development	Accountable for the proper design and development of the software components	
Head of architecture	Designing architecture in an optimal way	
Culture, Ethics and Behaviour Enabler		
Reference	**Contribution to Response to Scenario**	
Testing is performed on all appropriate levels	Users and developers co-operate in testing the software components.	
Information Enabler		
Reference	**Contribution to Response to Scenario**	
Architecture model	Target architecture model	
Design specifications	Clarifying the needs of the users	
Quality assurance plan (test plan and procedures)	Defining the steps to take in order to assure quality	
Maintenance plan	Planning the maintenance of the software	
Services, Infrastructure and Applications Enabler		
Reference	**Contribution to Response to Scenario**	
Integrated development environment (IDE)	Facilitating development and consisting of a source code editor, build automation tools and a debugger	
Knowledge repositories	Sharing and co-ordinating knowledge regarding development activities	
People, Skills and Competencies Enabler		
Reference	**Contribution to Response to Scenario**	
Architecture skills	Develop efficient and effective architecture aligned to the business requirements	
Technical skills	Designing and developing the proper software components	

D.10. Scenario 10: Business Ownership of IT		
Risk Scenario Category	Business ownership of IT	
Principles, Policies and Frameworks Enabler		
Reference	**Contribution to Response to Scenario**	
Enterprise governance guiding principles	Involving business and IT	
Reporting and communication principles	Clarifying the means of communication	
Process Enabler		
Reference	**Title**	**Management Practice**
EDM01.01	Evaluate the governance system.	Continually identify and engage with the enterprise's stakeholders, document an understanding of the requirements, and make a judgement on the current and future design of governance of enterprise IT.
EDM01.02	Direct the governance system.	Inform leaders and obtain their support, buy-in and commitment. Guide the structures, processes and practices for the governance of IT in line with agreed-on governance design principles, decision-making models and authority levels. Define the information required for informed decision making.
EDM01.03	Monitor the governance system.	Monitor the effectiveness and performance of the enterprise's governance of IT. Assess whether the governance system and implemented mechanisms (including structures, principles and processes) are operating effectively and provide appropriate oversight of IT.
APO01.04	Communicate management objectives and direction.	Communicate awareness and understanding of IT objectives and direction to appropriate stakeholders and users throughout the enterprise.
APO02.01	Understand enterprise direction.	Consider the current enterprise environment and business processes, as well as the enterprise strategy and future objectives. Consider also the external environment of the enterprise (industry drivers, relevant regulations, basis for competition).
APO05.06	Manage benefits achievement.	Monitor the benefits of providing and maintaining appropriate IT services and capabilities, based on the agreed-on and current business case.
APO09.03	Define and prepare service agreements.	Define and prepare service agreements based on the options in the service catalogues. Include internal operational agreements.
APO09.04	Monitor and report service levels.	Monitor service levels, report on achievements and identify trends. Provide the appropriate management information to aid performance management.
BAI01.03	Manage stakeholder engagement.	Manage stakeholder engagement to ensure an active exchange of accurate, consistent and timely information that reaches all relevant stakeholders. This includes planning, identifying and engaging stakeholders and managing their expectations.
BAI02.01	Define and maintain business functional and technical requirements.	Based on the business case, identify, prioritise, specify and agree on business information, functional, technical and control requirements covering the scope/understanding of all initiatives required to achieve the expected outcomes of the proposed IT-enabled business solution.
Organisational Structures Enabler		
Reference	**Contribution to Response to Scenario**	
Programme and project management office (PMO)	Provide a common methodology, used by business and IT, to define proper requirements.	
Value management office (VMO)	VMO, or similar role/collection of activities, needs to provide a common methodology, used by business and IT, to assess opportunities in terms of value for the enterprise.	
Strategy (IT executive) committee	Key structure that should take accountability over IT and business co-operation	
Board	Accountable for the governance framework setting and maintenance	
Culture, Ethics and Behaviour Enabler		
Reference	**Contribution to Response to Scenario**	
Business and IT work together as partners	Business takes into account the difficulties IT faces, IT learns the business issues	
Information Enabler		
Reference	**Contribution to Response to Scenario**	
IT strategy	Aligning your IT plans with business objectives will lead to a more efficient accountability of the business over IT.	
Authority levels	Clarifying the decision-making responsibilities	
SLAs	Service level agreements	

D.10. Scenario 10: Business Ownership of IT *(cont.)*	
Services, Infrastructure and Applications Enabler	
Reference	**Contribution to Response to Scenario**
N/A	N/A
People, Skills and Competencies Enabler	
Reference	**Contribution to Response to Scenario**
Relationship management skills	IT should have the proper skills to build relations with relevant business stakeholders
IT-related skills/ affinity	Business representatives should be trained/selected based on a minimal required affinity with IT

D.11. Scenario 11: Suppliers	
Risk Scenario Category	Suppliers (selection, performance, contractual compliance, termination of service and transfer)

Principles, Policies and Frameworks Enabler

Reference	Contribution to Response to Scenario
Procurement policy	Providing a set approach to selecting suppliers including the acceptance criteria for terms of business
Architecture principles	Architecture principles define the underlying general rules and guidelines for the use and deployment of all IT resources and assets across the enterprise.
Information security policy	Defines technical limitations on sharing and using information.

Process Enabler

Reference	Title	Management Practice
APO10.02	Select suppliers.	Select suppliers according to a fair and formal practice to ensure a viable best fit based on specified requirements. Requirements should be optimised with input from potential suppliers.
APO10.03	Manage supplier relationships and contracts.	Formalise and manage the supplier relationship for each supplier. Manage, maintain and monitor contracts and service delivery. Ensure that new or changed contracts conform to enterprise standards and legal and regulatory requirements. Deal with contractual disputes.
APO10.04	Manage supplier risk.	Identify and manage risk relating to suppliers' ability to continually provide secure, efficient and effective service delivery.
APO10.05	Monitor supplier performance and compliance.	Periodically review the overall performance of suppliers, compliance to contract requirements, and value for money, and address identified issues.

Organisational Structures Enabler

Reference	Contribution to Response to Scenario
Legal group	Review of proposed terms of business
Business process owner	Setting requirements, performance indicators and ensure proper expectations are incorporated in the contracts
Procurement department	Provide the support and approach to efficiently engage with suppliers.
CIO	Accountable for managing suppliers

Culture, Ethics and Behaviour Enabler

Reference	Contribution to Response to Scenario
Respect procurement procedures	Additional effort is required to ensure minimal protection regarding suppliers.
Transparent and participative culture is an important focus point.	To optimise the outcome of the vendor relationship

Information Enabler

Reference	Contribution to Response to Scenario
Service requirements	Knowing what you want allows for a reasonable position for negotiation.
IT strategy	Defining boundaries and enterprise objectives to take into account when negotiating contracts
Supplier catalogue	A structured presentation of known suppliers, including previous performance
SLAs	Service level agreements

Services, Infrastructure and Applications Enabler

Reference	Contribution to Response to Scenario
Vendor management system	Set up a system to keep track of the evolution of exposure to risk during the entire process, from selection until termination of service.

People, Skills and Competencies Enabler

Reference	Contribution to Response to Scenario
Negotiation skills	Ensure minimal requirements are supported.
Litigation skills	Once prosecution is initiated, the proper skills are required to minimize legal impact.
Legal analysis skills	Support co-operation with supplier.

D.12. Scenario 12: Regulatory Compliance		
Risk Scenario Category	Regulatory compliance	
Principles, Policies and Frameworks Enabler		
Reference	**Contribution to Response to Scenario**	
Domain specific policies	Policies such as privacy and health and safety	
Compliance policy	Guiding the identification of external compliance requirements	
Process Enabler		
Reference	**Title**	**Management Practice**
MEA03.01	Identify external compliance requirements.	On a continuous basis, identify and monitor for changes in local and international laws, regulations and other external requirements that must be complied with from an IT perspective.
MEA03.02	Optimise response to external requirements.	Review and adjust policies, principles, standards, procedures and methodologies to ensure that legal, regulatory and contractual requirements are addressed and communicated. Consider industry standards, codes of good practice, and best practice guidance for adoption and adaptation.
MEA03.03	Confirm external compliance.	Confirm compliance of policies, principles, standards, procedures and methodologies with legal, regulatory and contractual requirements.
Organisational Structures Enabler		
Reference	**Contribution to Response to Scenario**	
Privacy officer	Monitor impacts of laws and make sure privacy directives are met.	
Compliance department	Provides guidance on legal, regulatory and contractual compliance. Tracks new and changing regulations.	
Legal group	Legal support during analysis and litigation	
Culture, Ethics and Behaviour Enabler		
Reference	**Contribution to Response to Scenario**	
Risk- and compliance-aware culture is present throughout the enterprise including the proactive identification and escalation of risk.	All members of the enterprise are empowered to facilitate regulatory compliance.	
Compliance is embedded in daily operations.	All members of the enterprise are empowered to facilitate regulatory compliance.	
Information Enabler		
Reference	**Contribution to Response to Scenario**	
Risk appetite/tolerance	Balancing compliance with enterprises risk appetite/tolerance	
Assurance reports	For example, SAS 70	
Internal control framework	Optimise the efficiency of internal control.	
Analysis of new legal and regulatory compliance requirements	Regulations imposed by government needs to be analysed.	
Services, Infrastructure and Applications Enabler		
Reference	**Contribution to Response to Scenario**	
Regulatory databases	Facilitating the follow-up of compliance requirements	
GRC tools	Overview of controls and practices to assure compliance	
People, Skills and Competencies Enabler		
Reference	**Contribution to Response to Scenario**	
Litigation skills	Once prosecution is initiated, the proper skills are required to minimize legal impact.	
Legal analysis skills	Understand expectations of local regulator.	
Contingency planning skills	Maintain options for continuous service.	
Internal control	Assess compliance with relevant regulations.	

D.13. Scenario 13: Geopolitical	
Risk Scenario Category	Geopolitical

Principles, Policies and Frameworks Enabler	
Reference	**Contribution to Response to Scenario**
Safe harbour principle	Safe harbour agreements reduce the likelihood of interference.

Process Enabler		
Reference	**Title**	**Management Practice**
DSS04.02	Maintain a continuity strategy.	Evaluate business continuity management options and choose a cost-effective and viable continuity strategy that will ensure enterprise recovery and continuity in the face of a disaster or other major incident or disruption.
MEA03.01	Identify external compliance requirements.	On a continuous basis, identify and monitor for changes in local and international laws, regulations and other external requirements that must be complied with from an IT perspective.
MEA03.02	Optimise response to external requirements.	Review and adjust policies, principles, standards, procedures and methodologies to ensure that legal, regulatory and contractual requirements are addressed and communicated. Consider industry standards, codes of good practice, and best practice guidance for adoption and adaptation.

Organisational Structures Enabler	
Reference	**Contribution to Response to Scenario**
Privacy officer	Monitor impacts of laws and make sure privacy directives are met.
Regulatory compliance department	Guidance on legal, regulatory and contractual compliance
Legal group	Legal support during analysis and litigation
Business continuity/disaster recovery plan	Maintain options for continuous service.

Culture, Ethics and Behaviour Enabler	
Reference	**Contribution to Response to Scenario**
Controlled growth and expansion	Ensure the regulations and external requirements are integrated

Information Enabler	
Reference	**Contribution to Response to Scenario**
Analysis of new regulations	Regulations imposed by local government need to be analysed.

Services, Infrastructure and Applications Enabler	
Reference	**Contribution to Response to Scenario**
External legal services	Gain advice on new regulations from local governments and the impact they have on the enterprise.

People, Skills and Competencies Enabler	
Reference	**Contribution to Response to Scenario**
Litigation skills	Once prosecution is initiated, the proper skills are required to minimize legal impact.
Legal analysis skills	Understand expectations of local regulator.
Contingency planning skills	Maintain options for continuous service.

D.14. Scenario 14: Infrastructure Theft or Destruction	
Risk Scenario Category	Infrastructure theft or destruction (outside the enterprise)

Principles, Policies and Frameworks Enabler	
Reference	**Contribution to Response to Scenario**
Physical and environmental information security policy	Restricting physical access to infrastructure in order to prevent destruction
Business continuity and disaster recovery policy	Validate recoverability of information, services, applications and infrastructure.

Process Enabler		
Reference	**Title**	**Management Practice**
DSS01.04	Manage the environment.	Maintain measures for protection against environmental factors. Install specialised equipment and devices to monitor and control the environment.
DSS01.05	Manage facilities.	Manage facilities, including power and communications equipment, in line with laws and regulations, technical and business requirements, vendor specifications, and health and safety guidelines.
DSS05.05	Manage physical access to IT assets.	Define and implement procedures to grant, limit and revoke access to premises, buildings and areas according to business needs, including emergencies. Access to premises, buildings and areas should be justified, authorised, logged and monitored. This should apply to all persons entering the premises, including staff, temporary staff, clients, vendors, visitors or any other third party.

Organisational Structures Enabler	
Reference	**Contribution to Response to Scenario**
Information security manager	Implementation of security measures
Head of IT operations	Responding to infrastructure theft and destruction

Culture, Ethics and Behaviour Enabler	
Reference	**Contribution to Response to Scenario**
Information security is practiced in daily operations.	To prevent unauthorised physical access
People respect the importance of information security policies and principles.	To prevent unauthorised physical access
Stakeholders are aware of how to identify and respond to threats to the enterprise.	To minimise impact of infrastructure theft and destruction

Information Enabler	
Reference	**Contribution to Response to Scenario**
Access requests	Monitors access to facilities
Access logs	Integrated reporting on access
Facilities assessments reports	Enterprise is aware of state and risk of facilities.

Services, Infrastructure and Applications Enabler	
Reference	**Contribution to Response to Scenario**
Access control	To prevent unauthorised physical access
Alarm and monitoring security system	To prevent unauthorised physical access

People, Skills and Competencies Enabler	
Reference	**Contribution to Response to Scenario**
Information security skills	Preventing and reducing the impact of infrastructure theft and destruction

D.15. Scenario 15: Malware	
Risk Scenario Category	Malware

Principles, Policies and Frameworks Enabler	
Reference	**Contribution to Response to Scenario**
Information security policy	Outlines information security arrangements within the enterprise.
Malicious software prevention policy	Details the preventive, detective and corrective measures in place across the enterprise to protect information systems and technology from malware.
Architecture principles	Information security requirements are embedded within the enterprise architecture and translated into a formal information security architecture.
Business continuity and disaster recovery policy	Validate recoverability of information, services, applications and infrastructure.

Process Enabler		
Reference	**Title**	**Management Practice**
APO01.03	Maintain the enablers of the management system.	Maintain the enablers of the management system and control environment for enterprise IT, and ensure that they are integrated and aligned with the enterprise's governance and management philosophy and operating style. These enablers include the clear communication of expectations/requirements. The management system should encourage cross-divisional co-operation and teamwork, promote compliance and continuous improvement, and handle process deviations (including failure).
APO01.08	Maintain compliance with policies and procedures.	Put in place procedures to maintain compliance with and performance measurement of policies and other enablers of the control framework, and enforce the consequences of non-compliance or inadequate performance. Track trends and performance and consider these in the future design and improvement of the control framework.
DSS05.01	Protect against malware.	Implement and maintain preventive, detective and corrective measures in place (especially up-to-date security patches and virus control) across the enterprise to protect information systems and technology from malware (e.g., viruses, worms, spyware, spam).
DSS05.07	Monitor the infrastructure for security-related events.	Using intrusion detection tools, monitor the infrastructure for unauthorised access and ensure that any events are integrated with general event monitoring and incident management.

Organisational Structures Enabler	
Reference	**Contribution to Response to Scenario**
Information security manager	Implementation of security measures
Head of IT operations	Leading the management response team to restore service in a timely fashion

Culture, Ethics and Behaviour Enabler	
Reference	**Contribution to Response to Scenario**
Information security is practiced in daily operations.	To prevent the installation of malware
People respect the importance of information security policies and principles.	To prevent the installation of malware
Stakeholders are aware of how to identify and respond to threats to the enterprise.	To minimise impact of the installation of malware
Awareness and training regarding malware, email and Internet usage.	To prevent the installation of malware

Information Enabler	
Reference	**Contribution to Response to Scenario**
Threat information	Intelligence regarding types of attacks
Monitoring reports	Identification of attack attempts, threat events, etc.

Services, Infrastructure and Applications Enabler	
Reference	**Contribution to Response to Scenario**
Firewall	Protection against malware
SIEM	Security information and event management
Malicious software protection tools	Protection against malware
Monitoring and alert services	Be notified in time of potential threats

D.15. Scenario 15: Malware *(cont.)*	
People, Skills and Competencies Enabler	
Reference	**Contribution to Response to Scenario**
Information security skills	Preventing and reducing the impact of malware
IT technical skills	Configuration of IT infrastructure, such as firewalls, etc., and the review of vendor products

D.16. Scenario 16: Logical Attacks

Risk Scenario Category	Logical attacks

Principles, Policies and Frameworks Enabler

Reference	Contribution to Response to Scenario
Information security policy	Outlines information security arrangements within the enterprise.
Technical security policies and procedure	Details the technical consequences of the information security policy.
Architecture principles	Information security requirements are embedded within the enterprise architecture and translated into a formal information security architecture.
Business continuity and disaster recovery policy	Validate recoverability of information, services, applications and infrastructure.

Process Enabler

Reference	Title	Management Practice
APO13.01	Establish and maintain an information security management system (ISMS).	Establish and maintain an ISMS that provides a standard, formal and continuous approach to security management for information, enabling secure technology and business processes that are aligned with business requirements and enterprise security management.
APO13.03	Monitor and review the ISMS.	Maintain and regularly communicate the need for, and benefits of, continuous information security improvement. Collect and analyse data about the ISMS, and improve the effectiveness of the ISMS. Correct non-conformities to prevent recurrence. Promote a culture of security and continual improvement.

Process Enabler

Reference	Title	Management Practice
BAI03.07	Prepare for solution testing.	Establish a test plan and required environments to test the individual and integrated solution components, including the business processes and supporting services, applications and infrastructure
DSS01.03	Monitor IT infrastructure.	Monitor the IT infrastructure and related events. Store sufficient chronological information in operations logs to enable the reconstruction, review and examination of the time sequences of operations and the other activities surrounding or supporting operations.
DSS04.03	Develop and implement a business continuity response.	Develop a business continuity plan (BCP) based on the strategy that documents the procedures and information in readiness for use in an incident to enable the enterprise to continue its critical activities.
DSS05.01	Protect against malware.	Implement and maintain preventive, detective and corrective measures in place (especially up-to-date security patches and virus control) across the enterprise to protect information systems and technology from malware (e.g., viruses, worms, spyware, spam).
DSS05.02	Manage network and connectivity security.	Use security measures and related management procedures to protect information over all methods of connectivity.
DSS05.07	Monitor the infrastructure for security-related events.	Using intrusion detection tools, monitor the infrastructure for unauthorised access and ensure that any events are integrated with general event monitoring and incident management.

Organisational Structures Enabler

Reference	Contribution to Response to Scenario
Information security manager	Implementation of security measures
Head of IT operations	Leading the management response team to restore service in a timely fashion
Service manager	In case attacks are successful, communicate with end user and help to manage the response.
Chief security architect	Design of security measures

Culture, Ethics and Behaviour Enabler

Reference	Contribution to Response to Scenario
Information security is practiced in daily operations.	To prevent logical attacks
People respect the importance of information security policies and principles.	To prevent logical attacks
Stakeholders are aware of how to identify and respond to threats to the enterprise.	To minimise impact of logical attacks

D.16. Scenario 16: Logical Attacks *(cont.)*	
Information Enabler	
Reference	**Contribution to Response to Scenario**
SLAs	Detailing the action to be undertaken in case of attack
Threat information	Intelligence regarding types of attacks
Monitoring reports	Identification of attack attempts, threat events, etc.
Services, Infrastructure and Applications Enabler	
Reference	**Contribution to Response to Scenario**
Firewall	Prevent successful logical attacks.
SIEM	Security information and event management
Network management tools/vulnerability scanners	Identifying weaknesses
Monitoring and alert services	Be notified in time of potential threats
People, Skills and Competencies Enabler	
Reference	**Contribution to Response to Scenario**
Information security skills	Preventing and reducing the impact of logical attacks
IT technical skills	Configuration of IT infrastructure such as firewalls, critical network components, etc.

D.17. Scenario 17: Industrial Action		
Risk Scenario Category	Industrial action	
Principles, Policies and Frameworks Enabler		
Reference	**Contribution to Response to Scenario**	
HR policy	Define rights and obligations of all staff, detailing acceptable and unacceptable behaviour by the employees, and in doing so managing the risk that is linked to human behaviour.	
Vendor management policy	Define backup or emergency service delivery options.	
Process Enabler		
Reference	**Title**	**Management Practice**
APO01.01	Define the organisational structure.	Establish an internal and extended organisational structure that reflects business needs and IT priorities. Put in place the required management structures (e.g., committees) that enable management decision making to take place in the most effective and efficient manner.
APO07.01	Maintain adequate and appropriate staffing.	Evaluate staffing requirements on a regular basis or on major changes to the enterprise or operational or IT environments to ensure that the enterprise has sufficient human resources to support enterprise goals and objectives. Staffing includes both internal and external resources.
APO07.02	Identify key IT personnel.	Identify key IT personnel while minimising reliance on a single individual performing a critical job function through knowledge capture (documentation), knowledge sharing, succession planning and staff backup.
APO07.05	Plan and track the usage of IT and business human resources.	Understand and track the current and future demand for business and IT human resources with responsibilities for enterprise IT. Identify shortfalls and provide input into sourcing plans, enterprise and IT recruitment processes sourcing plans, and business and IT recruitment processes.
Organisational Structures Enabler		
Reference	**Contribution to Response to Scenario**	
Head of HR	Responsible for establishing expectations from and towards staff	
Legal group	Support initial contracting and prosecution in case of misuse.	
Board	Accountable for the well-functioning of the enterprise, top-level organisational structure for stakeholder communication	
Business executive	Facilitating two-way communication	
Culture, Ethics and Behaviour Enabler		
Reference	**Contribution to Response to Scenario**	
Transparent and participative culture is an important focus point.	To prevent industrial action from occurring	
Information Enabler		
Reference	**Contribution to Response to Scenario**	
Contract agreement with staff	Clear definition of responsibilities, rights and obligations for all individual staff	
Supplier contracts	Clear definition of responsibilities, rights and obligations for specific arrangements with vendors	
Knowledge repositories	Minimizing the effect of partial unavailability of resources by sharing knowledge regarding processes, technology, etc.	
Resource shortfall analysis	Clear analysis of critical level of resources	
Services, Infrastructure and Applications Enabler		
Reference	**Contribution to Response to Scenario**	
Third-party backup services	Temporary support in case of industrial action	
People, Skills and Competencies Enabler		
Reference	**Contribution to Response to Scenario**	
HR skills	Management of skills and competencies	
Negotiation skills	Facilitate the maximal two-way communication and ensure that minimal operational requirements are met.	
Litigation skills	Once prosecution is initiated, the proper skills are required to defend the interests of the enterprise.	

D.18. Scenario 18: Environmental	
Risk Scenario Category	Environmental

Principles, Policies and Frameworks Enabler

Reference	Contribution to Response to Scenario
Ethics policy	Environmental awareness should be part of the overall ethics policy.
Vendor management policy	Environmental awareness should be included in all contracts and agreements with vendors.
Rules of behaviour (acceptable use)	Users should be made aware of their individual impact in this regard.

Process Enabler

Reference	Title	Management Practice
APO02.03	Define the target IT capabilities.	Define the target business and IT capabilities and required IT services. This should be based on the understanding of the enterprise environment and requirements; the assessment of the current business process and IT environment and issues; and consideration of reference standards, best practices and validated emerging technologies or innovation proposals.
APO04.03	Monitor and scan the technology environment.	Perform systematic monitoring and scanning of the enterprise's external environment to identify emerging technologies that have the potential to create value (e.g., by realising the enterprise strategy, optimising costs, avoiding obsolescence, and better enabling enterprise and IT processes). Monitor the marketplace, competitive landscape, industry sectors, and legal and regulatory trends to be able to analyse emerging technologies or innovation ideas in the enterprise context.
BAI03.04	Procure solution components.	Procure solution components based on the acquisition plan in accordance with requirements and detailed designs, architecture principles and standards, and the enterprise's overall procurement and contract procedures, QA requirements, and approval standards. Ensure that all legal and contractual requirements are identified and addressed by the supplier.
DSS01.04	Manage the environment.	Maintain measures for protection against environmental factors. Install specialised equipment and devices to monitor and control the environment.
DSS01.05	Manage facilities.	Manage facilities, including power and communications equipment, in line with laws and regulations, technical and business requirements, vendor specifications, and health and safety guidelines.

Organisational Structures Enabler

Reference	Contribution to Response to Scenario
Head of IT operations	Responsible for managing the IT environment and facilities
Head architect	Design of environmental friendly measures

Culture, Ethics and Behaviour Enabler

Reference	Contribution to Response to Scenario
A clearly defined structure for ethical responsibility and a culture that promotes specific accountability is developed and supported.	People are involved and aware of the consequences of environmental issues and are empowered to handle according to ethical guidelines.

Information Enabler

Reference	Contribution to Response to Scenario
IT strategy	Environmental awareness should be part of the IT strategy.
Asset register	To assess the environmental impact of the used technology

Services, Infrastructure and Applications Enabler

Reference	Contribution to Response to Scenario
CMDB	Configuration management database assists in identifying areas for improvement

People, Skills and Competencies Enabler

Reference	Contribution to Response to Scenario
Architecture development	Architectural development can assist to reduce the environmental impact of technology.
System ergonomics	Streamlining and optimising used technology

D.19. Scenario 19: Acts of Nature	
Risk Scenario Category	Acts of nature
Principles, Policies and Frameworks Enabler	
Reference	**Contribution to Response to Scenario**
Backup policy	Backups are available.
Business continuity and disaster recovery policy	Validate recoverability of data.

Process Enabler			
Reference	**Title**	**Management Practice**	
DSS01.04	Manage the environment.	Maintain measures for protection against environmental factors. Install specialised equipment and devices to monitor and control the environment.	
DSS01.05	Manage facilities.	Manage facilities, including power and communications equipment, in line with laws and regulations, technical and business requirements, vendor specifications, and health and safety guidelines.	
DSS04.03	Develop and implement a business continuity response.	Develop a business continuity plan (BCP) based on the strategy that documents the procedures and information in readiness for use in an incident to enable the enterprise to continue its critical activities.	
DSS04.04	Exercise, test and review the BCP.	Test the continuity arrangements on a regular basis to exercise the recovery plans against predetermined outcomes and to allow innovative solutions to be developed and help to verify over time that the plan will work as anticipated.	
DSS05.05	Manage physical access to IT assets.	Define and implement procedures to grant, limit and revoke access to premises, buildings and areas according to business needs, including emergencies. Access to premises, buildings and areas should be justified, authorised, logged and monitored. This should apply to all persons entering the premises, including staff, temporary staff, clients, vendors, visitors or any other third party.	

Organisational Structures Enabler	
Reference	**Contribution to Response to Scenario**
Business continuity manager	Accountable for BCP plan
Head IT operations	Responsible for managing the IT environment and facilities
CIO	Responsible for developing and implementing a business continuity response
Business process owners	Responsible for developing and implementing a business continuity response

Culture, Ethics and Behaviour Enabler	
Reference	**Contribution to Response to Scenario**
Stakeholders are aware of how to identify and respond to threats.	People are involved and aware of how to react when an incident occurs.
Business management engages in continuous cross-functional collaboration to allow for efficient and effective business continuity programmes.	Business is committed and proactively contributes to risk mitigation.

Information Enabler	
Reference	**Contribution to Response to Scenario**
Insurance policy reports	Insurance in case of acts of nature is available.
Facilities assessments reports	Enterprise is aware of state and risk of facilities.
Incident response actions and communications	People are aware of how to react when an incident occurs.

Services, Infrastructure and Applications Enabler	
Reference	**Contribution to Response to Scenario**
Monitoring and alert services	Be notified in time of potential threats

People, Skills and Competencies Enabler	
Reference	**Contribution to Response to Scenario**
Information risk management	Identify and formulate response to information risk related to acts of nature
Technical understanding	Technical expertise regarding specific and relevant acts of nature

D.20. Scenario 20: Innovation		
Risk Scenario Category	Innovation	
Principles, Policies and Frameworks Enabler		
Reference	**Contribution to Response to Scenario**	
Architecture principles	Architecture principles define the underlying general rules and guidelines for the use and deployment of all IT resources and assets across the enterprise.	
Process Enabler		
Reference	**Title**	**Management Practice**
APO02.01	Understand enterprise direction.	Consider the current enterprise environment and business processes, as well as the enterprise strategy and future objectives. Consider also the external environment of the enterprise (industry drivers, relevant regulations, basis for competition).
APO02.03	Define the target IT capabilities.	Define the target business and IT capabilities and required IT services. This should be based on the understanding of the enterprise environment and requirements; the assessment of the current business process and IT environment and issues; and consideration of reference standards, best practices and validated emerging technologies or innovation proposals.
APO03.01	Develop the enterprise architecture vision.	The architecture vision provides a first-cut, high-level description of the baseline and target architectures, covering the business, information, data, application and technology domains. The architecture vision provides the sponsor with a key tool to sell the benefits of the proposed capability to stakeholders within the enterprise. The architecture vision describes how the new capability will meet enterprise goals and strategic objectives and address stakeholder concerns when implemented.
APO04.01	Create an environment conducive to innovation.	Create an environment that is conducive to innovation, considering issues such as culture, reward, collaboration, technology forums, and mechanisms to promote and capture employee ideas.
APO04.02	Maintain an understanding of the enterprise environment.	Work with relevant stakeholders to understand their challenges. Maintain an adequate understanding of enterprise strategy and the competitive environment or other constraints so that opportunities enabled by new technologies can be identified.
APO04.03	Monitor and scan the technology environment.	Perform systematic monitoring and scanning of the enterprise's external environment to identify emerging technologies that have the potential to create value (e.g., by realising the enterprise strategy, optimising costs, avoiding obsolescence, and better enabling enterprise and IT processes). Monitor the marketplace, competitive landscape, industry sectors, and legal and regulatory trends to be able to analyse emerging technologies or innovation ideas in the enterprise context.
APO04.04	Assess the potential of emerging technologies and innovation ideas.	Analyse identified emerging technologies and/or other IT innovation suggestions. Work with stakeholders to validate assumptions on the potential of new technologies and innovation.
APO04.05	Recommend appropriate further initiatives.	Evaluate and monitor the results of proof-of-concept initiatives and, if favourable, generate recommendations for further initiatives and gain stakeholder support
APO04.06	Monitor the implementation and use of innovation.	Monitor the implementation and use of emerging technologies and innovations during integration, adoption and for the full economic life cycle to ensure that the promised benefits are realised and to identify lessons learned.
Organisational Structures Enabler		
Reference	**Contribution to Response to Scenario**	
Board, CEO, strategy committee, CIO, innovation group		
Culture, Ethics and Behaviour Enabler		
Reference	**Contribution to Response to Scenario**	
Willingness to take risk, support of senior management for innovation initiatives and 'failure is allowed' attitude		
Information Enabler		
Reference	**Contribution to Response to Scenario**	
Innovation plan, recognition program, evaluation of innovation initiatives		
Services, Infrastructure and Applications Enabler		
Reference	**Contribution to Response to Scenario**	
N/A	N/A	

D.20. Scenario 20: Innovation *(cont.)*	
People, Skills and Competencies Enabler	
Reference	**Contribution to Response to Scenario**
Leadership and communication	Clarify the rationale for the architecture and the potential consequences.
Architecture skills	Develop efficient and effective architecture aligned to the business requirements.

APPENDIX E
COMPARISON OF RISK IT WITH COBIT 5

This appendix contains a comparison between the relevant parts of the Risk IT guidance (both *The Risk IT Framework* and *The Risk IT Practitioner Guide*) and their COBIT 5 equivalents, either contained in the COBIT 5 framework, *COBIT 5: Enabling Processes* or *COBIT 5 for Risk*.

E.1 Comparison of *The Risk IT Framework* With COBIT 5				
The Risk IT Framework Chapter	Subsection	COBIT 5	*COBIT 5: Enabling Processes*	*COBIT 5 for Risk*
3. Risk IT principles: • Connect to business objectives • Align IT risk management with ERM • Balance cost/benefit of IT risk • Promote fair and open communication • Establish tone at the top • Function as part of daily activities		Risk optimisation is one of the three core components of the value objective of the enterprise.	N/A	The risk principles are included in the Principles, Policies and Frameworks enabler (section 2A, chapter 2). A more detailed description of risk principles is included in appendix B.1.
4. *The Risk IT Framework*		The IT risk-related processes are integrated in the COBIT 5 Process Reference Model		N/A
5. Essentials of risk governance	A. Risk appetite and tolerance	N/A	Risk appetite and risk tolerance is (partly) covered by process EDM03.	Risk appetite and tolerance are defined as examples of the Information enabler (section 2A, chapter 6). A more detailed description of risk capacity, risk appetite and risk tolerance is included in appendix B.5. This section also includes guidance on how to set and communicate these thresholds.
	B. Responsibilities and accountability for IT risk management	Responsibilities and accountability are part of the RACI chart in the process description.	Responsibilities and accountability are part of the RACI chart in the process description.	Responsibilities and accountability for IT risk management are defined as part of the Organisational Structures enabler (section 2A, chapter 4).
	C. Awareness and communication	Culture, Ethics and Behaviour are now included as a separate enabler in the COBIT 5 framework, including both awareness and communication as practices.	Risk awareness and communication is (partly) covered by process EDM03.	Awareness and communication are defined as part of the Culture, Ethics and Behaviour enabler (section 2A, chapter 5).
	D. Risk culture	Culture, Ethics and Behaviour are now included as a separate enabler in the COBIT 5 framework.	Risk culture is covered by practice EDM03.02.	The Culture, Ethics and Behaviour enabler is described in section 2A, chapter 5. A more detailed description of desired behaviours and how to achieve them is included in appendix B.4.

The Risk IT Framework chapter	Subsection	COBIT 5	COBIT 5: Enabling Processes	COBIT 5 for Risk
		E.1 Comparison of *The Risk IT Framework* With COBIT 5 *(cont.)*		
6. Essentials of risk evaluation	A. Describing business impact	Principle 1: Meeting stakeholder needs, covers the COBIT 5 goals cascade. It is the mechanism to translate stakeholder needs into specific, actionable and customised enterprise goals, IT-related goals and enabler goals.	Business impact is covered by process EDM03.	Business impact is included in the elaboration of the Processes enabler in appendix B.2.
	B. IT risk scenarios	N/A	IT risk scenarios are covered in practices APO12.03/04.	All information regarding IT risk scenarios is included in section 2B, chapter 2. A more detailed description of the IT risk scenarios and how to respond to them using COBIT 5 is included in appendix D.
7. Essentials of risk response	A. Key risk indicators	N/A	KRIs are (partly) covered by process EDM03.	KRIs are defined as examples of the Information enabler (section 2A, chapter 6). A more detailed description of KRIs is included in appendix B.5.
	B. Risk response selection and prioritisation	N/A	Risk response selection and prioritisation is covered in practice APO12.06.	All information regarding risk response is included in section 2B, chapter 5.
8. Risk and opportunity management using COBIT, Val IT and Risk IT		Principle 3: Applying a single integrated framework, implies that all knowledge covered in previous ISACA frameworks has been integrated in COBIT 5.	N/A	A more detailed description of the IT risk scenarios and how to respond to them using COBIT 5 is included in appendix D.
12. *The Risk IT Framework*	RG1	IT risk-related processes are integrated in the COBIT 5 process reference model.	EDM03, APO12	Further process details are included in the processes enabler detailed description in appendix B.2, Enabler: Processes.
	RG2		EDM03, EDM04, APO07	
	RG3		EDM01, EDM03	
	RE1		APO12.01	
	RE2		APO12.02	
	RE3		APO12.02, APO12.03	
	RR1		APO12.04	
	RR2		APO12.05	
	RR3		APO12.06	
Appendix 2. High-level comparison with other risk management frameworks		N/A	N/A	Appendix C

E.2 Comparison of The Risk IT Practitioner Guide With COBIT 5				
The Risk IT Practitioner Guide chapter	**Subsection**	**COBIT 5**	**COBIT 5: Enabling Processes**	**COBIT 5 for Risk**
1. Defining a risk universe and scoping risk management		The COBIT 5 framework is used to establish the boundaries of the GEIT universe for the enterprise, i.e., all aspects of the GEIT universe need to be considered from a governance and management perspective.	Risk universe and scoping risk is (partly) covered by process EDM03.	COBIT 5 for Risk guides enterprises in the use of COBIT to consider and address GEIT from a risk governance and management perspective and supports enterprise risk professionals in their use of COBIT in performing their activities.
2. Risk appetite and risk tolerance		N/A	Risk appetite and risk tolerance is (partly) covered by process EDM03.	Risk appetite and tolerance are defined as examples of the Information enabler (section 2A, chapter 6). More detailed descriptions of risk capacity, risk appetite and risk tolerance are included in appendix B.5. This appendix also includes guidance on how to set and communicate these thresholds.
3. Risk awareness, communication and reporting	A. Risk awareness and communication	Culture, Ethics and Behaviour are now included as a separate enabler in the COBIT 5 framework.	Risk awareness and communication is (partly) covered by process EDM03.	Awareness and communication are defined as part of the Culture, Ethics and Behaviour enabler (section 2A, chapter 5).
	B. Key risk indicators and risk reporting	N/A	KRIs and risk reporting is (partly) covered by process EDM03.	KRIs are defined as examples of the Information enabler (section 2A, chapter 6). A more detailed description of KRIs is included in appendix B.5.
	C. Risk profile	N/A	Risk profile is covered in practice APO12.03.	Risk profiles are defined as examples of the Information enabler (section 2A, chapter 6). A more detailed description of risk profiles is included in appendix B.5.
	D. Risk aggregation	N/A	Risk aggregation is (partly) covered by process EDM03.	Risk aggregation is explained in section 2B, chapter 4.
	E. Risk culture	Culture, Ethics and Behaviour are now included as a separate enabler in the COBIT 5 framework.	Risk culture is covered by practice EDM03.02.	The Culture, Ethics and Behaviour enabler is described in section 2A, chapter 5. A more detailed description of desired behaviours and how to achieve them is included in appendix B.4.

E.2 Comparison of The Risk IT Practitioner Guide With COBIT 5 *(cont.)*				
The Risk IT Practitioner Guide chapter	**Subsection**	**COBIT 5**	*COBIT 5: Enabling Processes*	*COBIT 5 for Risk*
4. Expressing and describing risk	A. Expressing impact in business terms	Principle 1: Meeting stakeholder needs, covers the COBIT 5 goals cascade. It is the mechanism to translate stakeholder needs into specific, actionable and customised enterprise goals, IT-related goals and enabler goals.	Business impact is covered by process EDM03.	Business impact is included in the elaboration of the Processes enabler in appendix B.2.
	B. Describing risk—expressing frequency	N/A	Expressing frequency is covered in practice APO12.02.	Expressing frequency is included in the elaboration of the processes enabler in appendix B.2. Guidance to the same level of detail or detailed examples as in the *Risk IT Practitioner Guide* is **not** included.
	C. Describing risk—expressing impact	N/A	Expressing impact is covered in practice APO12.02.	Expressing impact is included in the elaboration of the processes enabler in appendix B.2. Guidance to the same level of detail or detailed examples as in the *Risk IT Practitioner Guide* is **not** included.
	D. COBIT business goal comparison with other impact criteria	N/A	Business impact is covered by process EDM03.	Business impact is included in the elaboration of the processes enabler in appendix B.2. Guidance to the same level of detail as in the *Risk IT Practitioner Guide* is **not** included.
	E. Risk map	N/A	The use of a risk map is covered in practices APO12.03/04.	Risk maps are defined as examples of the Information enabler (section 2A, chapter 6). A more detailed description of the information item risk map is included in appendix B.5. Guidance to the same level of detail or detailed examples as in *The Risk IT Practitioner Guide* is **not** included.
	F. Risk register	N/A	The use of a risk register is covered in practices APO12.03/04.	Risk register is defined as part of the risk profile (section 2A, chapter 6) A more detailed description of risk profiles is included in appendix B.5. This section also contains a template for a risk register entry.

E.2 Comparison of The Risk IT Practitioner Guide With COBIT 5 *(cont.)*				
The Risk IT Practitioner Guide chapter	Subsection	COBIT 5	COBIT 5: Enabling Processes	*COBIT 5 for Risk*
5. Risk scenarios	A. Risk scenarios explained	N/A	N/A	All information regarding IT risk scenarios is included in section 2B, chapters 1 through 3. A more detailed description of the IT risk scenarios and how to respond to them using COBIT 5 enablers is included in appendix D.
	B. Example risk scenarios	N/A	N/A	Example risk scenarios are listed in section 2B, chapter 3. A more detailed description of the IT risk scenarios and how to respond to them using COBIT 5 enablers is included in appendix D.
	C. Capability risk factors in the risk analysis process	N/A	N/A	All information regarding IT risk scenarios is included in section 2B, chapters 1 and 2. A more detailed description of the IT risk scenarios and how to respond to them using COBIT 5 enablers is included in appendix D.
	D. Environmental risk factors in the risk analysis process	N/A	N/A	All information regarding IT risk scenarios is included in section 2B, chapters 1 and 2. A more detailed description of the IT risk scenarios and how to respond to them using COBIT 5 enablers is included in appendix D.
6. Risk response and prioritisation		N/A	Risk response covered by process APO12.	All information regarding risk response is included in section 2B, chapter 5.
7. A risk analysis workflow		N/A	Risk analysis workflow covered by process APO12.	N/A
8. Mitigation of IT risk using COBIT and Val IT		N/A	N/A	Appendix D contains a detailed description on how to respond to each of the identified risk scenario categories (20) using COBIT 5 enablers.

Page intentionally left blank

APPENDIX F
COMPREHENSIVE RISK SCENARIO TEMPLATE

This appendix contains a comprehensive template for the treatment of a risk scenario—from conception through response and monitoring—in support of the core risk management processes of an enterprise.

Risk Scenario Template	
Title	
Category High-level description of the scenario category	☐ 01-Portfolio establishment and maintenance ☐ 02-Programme/project life cycle management ☐ 03-IT investment decision making ☐ 04-IT expertise and skills ☐ 05-Staff operations ☐ 06-Information ☐ 07-Architecture ☐ 08-Infrastructure ☐ 09-Software ☐ 10-Ineffective business ownership of IT ☐ 11-Selection/performance of third-party suppliers ☐ 12-Regulatory compliance ☐ 13-Geopolitical ☐ 14-Infrastructure theft ☐ 15-Malware ☐ 16-Logical attacks ☐ 17-Industrial action ☐ 18-Environmental ☐ 19-Acts of nature ☐ 20-Innovation
Describe the risk/opportunity scenario, including a discussion of the negative and positive impact of the scenario. The description clarifies the threat/vulnerability type and includes the actors, events, assets and time issues.	
Threat Type The nature of the event—Is it malicious? If not, is it accidental or is it a failure of a well-defined process? Is it a natural event or is it an external requirement?	☐ **Malicious** ☐ **Accidental** ☐ **Error** ☐ **Failure** ☐ **Natural** ☐ **External requirement**
Actor Who generates the threat that exploits a vulnerability? Actors can be internal or external, human or non-human.	☐ **Internal** *actors are within the enterprise, e.g., staff, contractors.* ☐ **External** *actors include outsiders, competitors, regulators and the market.*
Event Is it disclosure (of confidential information), interruption (of a system or a project), theft or destruction? Action also includes ineffective design (of systems, processes, etc.), inappropriate use, changes in rules and regulations that materially impact a system, or ineffective execution of processes, e.g., change management procedures, acquisition procedures, project prioritisation processes.	☐ **Disclosure** ☐ **Interruption** ☐ **Modification** ☐ **Theft** ☐ **Destruction** ☐ **Ineffective design** ☐ **Ineffective execution** ☐ **Rules and regulations** ☐ **Inappropriate use**
Asset An asset is any item of value to the enterprise that can be affected and lead to business impact. (Assets and resources can be identical, e.g., IT hardware is an important resource because all IT applications use it, and, at the same time, it is an asset because it has a certain value to the enterprise.)	1. **Process**, e.g., modelled as COBIT 5 processes or business processes 2. **People and Skills** 3. **Organisational Structure** 4. **Physical Infrastructure** (facilities, equipment, etc.) 5. **IT Infrastructure** (including computing hardware, networks, middleware) 6. **Information** 7. **Applications**
Resource A resource is anything that helps to achieve a goal. (Assets and resources can be identical, e.g., IT hardware is an important resource because all IT applications use it, and at the same time, it is an asset because it has a certain value to the enterprise.)	1. **Process**, e.g., modelled as COBIT 5 processes or business processes 2. **People and Skills** 3. **Organisational Structure** 4. **Physical Infrastructure** (facilities, equipment, etc.) 5. **IT Infrastructure** (including computing hardware, networks, middleware) 6. **Information** 7. **Applications**

Risk Scenario Template *(cont.)*

Title	

Time	1. Timing of occurrence (critical, non-critical—Does the event occur at a critical moment?) 2. Duration (extended—The duration of the event, e.g., extended outage of a service or data centre 3. Detection (slow, moderate, instant) 4. Time Lag (immediate, delayed—Lag between the event and the consequence; Is there an immediate consequence [e.g., network failure, immediate downtime, or delayed consequence] or an incorrect IT architecture with accumulated high costs over a time span of several years?)

Risk Type
Describe the risk type. Include whether the risk type is primary or secondary, i.e., a higher or lower degree of fit.
Risk Types:
- **IT Benefit/Value Enablement:**
  Associated with [missed] opportunities to use technology to improve efficiency or effectiveness of business processes, or as an enabler for new business initiatives
  − Technology enabler for new business initiatives
  − Technology enabler for efficient operations
- **IT Programme and Project Delivery:**
  Associated with the contribution of IT to new or improved business solutions, usually in the form of projects and programmes as part of investment portfolios.
  − Project quality
  − Project relevance
  − Project overrun
- **IT Operations and Service Delivery:**
  Associated with all aspects of the business-as-usual performance of IT systems and services, which can bring destruction or reduction of value to the enterprise.
  − IT service interruptions
  − Security problems
  − Compliance issues

Risk Response
Describe how the enterprise will respond to the risk. The purpose of defining a risk response is to bring risk in line with the defined risk appetite and tolerance for the enterprise:
- Risk acceptance
- Risk sharing/transfer
- Risk mitigation
- Risk avoidance

Risk Mitigation Using COBIT 5 Enablers (see appendix D in COBIT 5 for Risk)
Describe how the enterprise will work to avoid the risk from materialising. For risk mitigation possibilities, use COBIT 5 management practices (enablers).
Provide the following information:
- Reference, title and description of one or more relevant enablers that can help to mitigate the risk
- The estimated effect that implementing this enabler will have on the frequency and impact of the risk. Use either a low, medium or high rating.
- Based on the two parameters, frequency and impact, indicate whether this enabler is 'essential' (key management practice to mitigate the risk). An enabler is considered essential if it has a high effect on reducing either impact or frequency of the scenario.

Management Practice Reference	Title	Management Practice	Effect on Frequency	Effect on Impact	Essential

Key Risk Indicators
Identify a number of metrics to detect and monitor the risk scenario and the risk response.